FREE Test Taking Tips DVD Offer

To help us better serve you, we have developed a Test Taking Tips DVD that we would like to give you for FREE. **This DVD covers world-class test taking tips that you can use to be even more successful when you are taking your test.**

All that we ask is that you email us your feedback about your study guide. Please let us know what you thought about it – whether that is good, bad or indifferent.

To get your **FREE Test Taking Tips DVD**, email freedvd@studyguideteam.com with "FREE DVD" in the subject line and the following information in the body of the email:

   a. The title of your study guide.

   b. Your product rating on a scale of 1-5, with 5 being the highest rating.

   c. Your feedback about the study guide. What did you think of it?

   d. Your full name and shipping address to send your free DVD.

If you have any questions or concerns, please don't hesitate to contact us at freedvd@studyguideteam.com.

Thanks again!

AP Biology Test Prep Book 2019 & 2020

AP Biology Review Book & Practice Test Questions

Test Prep Books

Table of Contents

Quick Overview

As you draw closer to taking your exam, effective preparation becomes more and more important. Thankfully, you have this study guide to help you get ready. Use this guide to help keep your studying on track and refer to it often.

This study guide contains several key sections that will help you be successful on your exam. The guide contains tips for what you should do the night before and the day of the test. Also included are test-taking tips. Knowing the right information is not always enough. Many well-prepared test takers struggle with exams. These tips will help equip you to accurately read, assess, and answer test questions.

A large part of the guide is devoted to showing you what content to expect on the exam and to helping you better understand that content. In this guide are practice test questions so that you can see how well you have grasped the content. Then, answer explanations are provided so that you can understand why you missed certain questions.

Don't try to cram the night before you take your exam. This is not a wise strategy for a few reasons. First, your retention of the information will be low. Your time would be better used by reviewing information you already know rather than trying to learn a lot of new information. Second, you will likely become stressed as you try to gain a large amount of knowledge in a short amount of time. Third, you will be depriving yourself of sleep. So be sure to go to bed at a reasonable time the night before. Being well-rested helps you focus and remain calm.

Be sure to eat a substantial breakfast the morning of the exam. If you are taking the exam in the afternoon, be sure to have a good lunch as well. Being hungry is distracting and can make it difficult to focus. You have hopefully spent lots of time preparing for the exam. Don't let an empty stomach get in the way of success!

When travelling to the testing center, leave earlier than needed. That way, you have a buffer in case you experience any delays. This will help you remain calm and will keep you from missing your appointment time at the testing center.

Be sure to pace yourself during the exam. Don't try to rush through the exam. There is no need to risk performing poorly on the exam just so you can leave the testing center early. Allow yourself to use all of the allotted time if needed.

Remain positive while taking the exam even if you feel like you are performing poorly. Thinking about the content you should have mastered will not help you perform better on the exam.

Once the exam is complete, take some time to relax. Even if you feel that you need to take the exam again, you will be well served by some down time before you begin studying again. It's often easier to convince yourself to study if you know that it will come with a reward!

Test-Taking Strategies

1. Predicting the Answer

When you feel confident in your preparation for a multiple-choice test, try predicting the answer before reading the answer choices. This is especially useful on questions that test objective factual knowledge. By predicting the answer before reading the available choices, you eliminate the possibility that you will be distracted or led astray by an incorrect answer choice. You will feel more confident in your selection if you read the question, predict the answer, and then find your prediction among the answer choices. After using this strategy, be sure to still read all of the answer choices carefully and completely. If you feel unprepared, you should not attempt to predict the answers. This would be a waste of time and an opportunity for your mind to wander in the wrong direction.

2. Reading the Whole Question

Too often, test takers scan a multiple-choice question, recognize a few familiar words, and immediately jump to the answer choices. Test authors are aware of this common impatience, and they will sometimes prey upon it. For instance, a test author might subtly turn the question into a negative, or he or she might redirect the focus of the question right at the end. The only way to avoid falling into these traps is to read the entirety of the question carefully before reading the answer choices.

3. Looking for Wrong Answers

Long and complicated multiple-choice questions can be intimidating. One way to simplify a difficult multiple-choice question is to eliminate all of the answer choices that are clearly wrong. In most sets of answers, there will be at least one selection that can be dismissed right away. If the test is administered on paper, the test taker could draw a line through it to indicate that it may be ignored; otherwise, the test taker will have to perform this operation mentally or on scratch paper. In either case, once the obviously incorrect answers have been eliminated, the remaining choices may be considered. Sometimes identifying the clearly wrong answers will give the test taker some information about the correct answer. For instance, if one of the remaining answer choices is a direct opposite of one of the eliminated answer choices, it may well be the correct answer. The opposite of obviously wrong is obviously right! Of course, this is not always the case. Some answers are obviously incorrect simply because they are irrelevant to the question being asked. Still, identifying and eliminating some incorrect answer choices is a good way to simplify a multiple-choice question.

4. Don't Overanalyze

Anxious test takers often overanalyze questions. When you are nervous, your brain will often run wild, causing you to make associations and discover clues that don't actually exist. If you feel that this may be a problem for you, do whatever you can to slow down during the test. Try taking a deep breath or counting to ten. As you read and consider the question, restrict yourself to the particular words used by the author. Avoid thought tangents about what the author *really* meant, or what he or she was *trying* to say. The only things that matter on a multiple-choice test are the words that are actually in the question. You must avoid reading too much into a multiple-choice question, or supposing that the writer meant something other than what he or she wrote.

5. No Need for Panic

It is wise to learn as many strategies as possible before taking a multiple-choice test, but it is likely that you will come across a few questions for which you simply don't know the answer. In this situation, avoid panicking. Because most multiple-choice tests include dozens of questions, the relative value of a single wrong answer is small. As much as possible, you should compartmentalize each question on a multiple-choice test. In other words, you should not allow your feelings about one question to affect your success on the others. When you find a question that you either don't understand or don't know how to answer, just take a deep breath and do your best. Read the entire question slowly and carefully. Try rephrasing the question a couple of different ways. Then, read all of the answer choices carefully. After eliminating obviously wrong answers, make a selection and move on to the next question.

6. Confusing Answer Choices

When working on a difficult multiple-choice question, there may be a tendency to focus on the answer choices that are the easiest to understand. Many people, whether consciously or not, gravitate to the answer choices that require the least concentration, knowledge, and memory. This is a mistake. When you come across an answer choice that is confusing, you should give it extra attention. A question might be confusing because you do not know the subject matter to which it refers. If this is the case, don't eliminate the answer before you have affirmatively settled on another. When you come across an answer choice of this type, set it aside as you look at the remaining choices. If you can confidently assert that one of the other choices is correct, you can leave the confusing answer aside. Otherwise, you will need to take a moment to try to better understand the confusing answer choice. Rephrasing is one way to tease out the sense of a confusing answer choice.

7. Your First Instinct

Many people struggle with multiple-choice tests because they overthink the questions. If you have studied sufficiently for the test, you should be prepared to trust your first instinct once you have carefully and completely read the question and all of the answer choices. There is a great deal of research suggesting that the mind can come to the correct conclusion very quickly once it has obtained all of the relevant information. At times, it may seem to you as if your intuition is working faster even than your reasoning mind. This may in fact be true. The knowledge you obtain while studying may be retrieved from your subconscious before you have a chance to work out the associations that support it. Verify your instinct by working out the reasons that it should be trusted.

8. Key Words

Many test takers struggle with multiple-choice questions because they have poor reading comprehension skills. Quickly reading and understanding a multiple-choice question requires a mixture of skill and experience. To help with this, try jotting down a few key words and phrases on a piece of scrap paper. Doing this concentrates the process of reading and forces the mind to weigh the relative importance of the question's parts. In selecting words and phrases to write down, the test taker thinks about the question more deeply and carefully. This is especially true for multiple-choice questions that are preceded by a long prompt.

9. Subtle Negatives

One of the oldest tricks in the multiple-choice test writer's book is to subtly reverse the meaning of a question with a word like *not* or *except*. If you are not paying attention to each word in the question, you can easily be led astray by this trick. For instance, a common question format is, "Which of the following is...?" Obviously, if the question instead is, "Which of the following is not...?," then the answer will be quite different. Even worse, the test makers are aware of the potential for this mistake and will include one answer choice that would be correct if the question were not negated or reversed. A test taker who misses the reversal will find what he or she believes to be a correct answer and will be so confident that he or she will fail to reread the question and discover the original error. The only way to avoid this is to practice a wide variety of multiple-choice questions and to pay close attention to each and every word.

10. Reading Every Answer Choice

It may seem obvious, but you should always read every one of the answer choices! Too many test takers fall into the habit of scanning the question and assuming that they understand the question because they recognize a few key words. From there, they pick the first answer choice that answers the question they believe they have read. Test takers who read all of the answer choices might discover that one of the latter answer choices is actually *more* correct. Moreover, reading all of the answer choices can remind you of facts related to the question that can help you arrive at the correct answer. Sometimes, a misstatement or incorrect detail in one of the latter answer choices will trigger your memory of the subject and will enable you to find the right answer. Failing to read all of the answer choices is like not reading all of the items on a restaurant menu: you might miss out on the perfect choice.

11. Spot the Hedges

One of the keys to success on multiple-choice tests is paying close attention to every word. This is never truer than with words like almost, most, some, and sometimes. These words are called "hedges" because they indicate that a statement is not totally true or not true in every place and time. An absolute statement will contain no hedges, but in many subjects, the answers are not always straightforward or absolute. There are always exceptions to the rules in these subjects. For this reason, you should favor those multiple-choice questions that contain hedging language. The presence of qualifying words indicates that the author is taking special care with his or her words, which is certainly important when composing the right answer. After all, there are many ways to be wrong, but there is only one way to be right! For this reason, it is wise to avoid answers that are absolute when taking a multiple-choice test. An absolute answer is one that says things are either all one way or all another. They often include words like *every*, *always*, *best*, and *never*. If you are taking a multiple-choice test in a subject that doesn't lend itself to absolute answers, be on your guard if you see any of these words.

12. Long Answers

In many subject areas, the answers are not simple. As already mentioned, the right answer often requires hedges. Another common feature of the answers to a complex or subjective question are qualifying clauses, which are groups of words that subtly modify the meaning of the sentence. If the question or answer choice describes a rule to which there are exceptions or the subject matter is complicated, ambiguous, or confusing, the correct answer will require many words in order to be expressed clearly and accurately. In essence, you should not be deterred by answer choices that seem excessively long. Oftentimes, the author of the text will not be able to write the correct answer without offering some qualifications and modifications. Your job is to read the answer choices thoroughly and completely and to select the one that most accurately and precisely answers the question.

13. Restating to Understand

Sometimes, a question on a multiple-choice test is difficult not because of what it asks but because of how it is written. If this is the case, restate the question or answer choice in different words. This process serves a couple of important purposes. First, it forces you to concentrate on the core of the question. In order to rephrase the question accurately, you have to understand it well. Rephrasing the question will concentrate your mind on the key words and ideas. Second, it will present the information to your mind in a fresh way. This process may trigger your memory and render some useful scrap of information picked up while studying.

14. True Statements

Sometimes an answer choice will be true in itself, but it does not answer the question. This is one of the main reasons why it is essential to read the question carefully and completely before proceeding to the answer choices. Too often, test takers skip ahead to the answer choices and look for true statements. Having found one of these, they are content to select it without reference to the question above. Obviously, this provides an easy way for test makers to play tricks. The savvy test taker will always read the entire question before turning to the answer choices. Then, having settled on a correct answer choice, he or she will refer to the original question and ensure that the selected answer is relevant. The mistake of choosing a correct-but-irrelevant answer choice is especially common on questions related to specific pieces of objective knowledge. A prepared test taker will have a wealth of factual knowledge at his or her disposal, and should not be careless in its application.

15. No Patterns

One of the more dangerous ideas that circulates about multiple-choice tests is that the correct answers tend to fall into patterns. These erroneous ideas range from a belief that B and C are the most common right answers, to the idea that an unprepared test-taker should answer "A-B-A-C-A-D-A-B-A." It cannot be emphasized enough that pattern-seeking of this type is exactly the WRONG way to approach a multiple-choice test. To begin with, it is highly unlikely that the test maker will plot the correct answers according to some predetermined pattern. The questions are scrambled and delivered in a random order. Furthermore, even if the test maker was following a pattern in the assignation of correct answers, there is no reason why the test taker would know which pattern he or she was using. Any attempt to discern a pattern in the answer choices is a waste of time and a distraction from the real work of taking the test. A test taker would be much better served by extra preparation before the test than by reliance on a pattern in the answers.

FREE DVD OFFER

Don't forget that doing well on your exam includes both understanding the test content and understanding how to use what you know to do well on the test. We offer a completely FREE Test Taking Tips DVD that covers world class test taking tips that you can use to be even more successful when you are taking your test.

All that we ask is that you email us your feedback about your study guide. To get your **FREE Test Taking Tips DVD**, email freedvd@studyguideteam.com with "FREE DVD" in the subject line and the following information in the body of the email:

- The title of your study guide.
- Your product rating on a scale of 1-5, with 5 being the highest rating.
- Your feedback about the study guide. What did you think of it?
- Your full name and shipping address to send your free DVD.

Introduction to the AP Biology Exam

Function of the Test

Like other Advanced Placement (AP) tests, the AP Biology test is offered by the College Board to high school students in the spring. Although students can register for and take an AP test without completing the related AP course, most test takers attempt the exam at the culmination of an AP course in the given AP subject, which typically lasts the duration of the academic year. A successful result on the test demonstrates mastery of college-level subject matter, and can be used by colleges to place students beyond entry-level courses and into more advanced courses. Favorable results and high scores can also be used on a student's college application to show academic success and the ability to handle college-level material.

In 2015, 223,479 students took the AP Biology test. Of these, 102,907 were in 12th grade, 84,157 were in 11th grade, and almost all of the rest were in 9th or 10th grades. Sixty percent of the students taking the test were female and 40 percent were male. The test and the AP program are offered nationwide, although not every high school may offer specific AP courses.

Test Administration

AP Biology exams are offered on a certain date in May each year and are mostly administered by schools that offer an AP Biology course. However, students can make arrangements with a school to take an AP exam even if they did not take the course at that particular school. All AP exams cost the same amount of money, with an additional fee added for exams administered outside of the U.S. and Canada. Schools can also add fees to cover their costs of administering the exams if they wish, but most offer the exams at the standard base rate.

Accommodations for students with documented disabilities include time extensions, large-type exams, large-block answer sheets, Braille devices, question readers, response writers, and more. Students seeking accommodations should contact the Disabilities Office of College Board Services.

Students may take an AP exam every time it is offered (i.e., once a year). Scores from all attempts will be reported in the score report after each test.

Test Format

The AP Biology exam comes in two sections. The first section contains sixty-three multiple-choice questions and six "grid-in" questions (requiring students to calculate an answer and enter it in a grid on the answer sheet), lasts ninety minutes, and comprises 50 percent of the score. This section is completed in pencil on a mechanized answer sheet. The second section contains two "long" free-response questions and six "short" free-response questions, lasts ninety minutes, and comprises the other 50 percent of the score. This section is generally answered by hand in pen.

The exam, like the course, is designed to cover four "big ideas." These ideas are as follows:

Concept	Approximate Share of Content
Big Idea 1- Evolution	25%
Big Idea 2- Biological systems and use of energy	25%
Big Idea 3- Living systems storage, retrieval, transmittal, and response to information	25%
Big Idea 4- Interaction of biological systems	25%

Scoring

On the multiple-choice section, students receive a raw score simply equal to one point for each correct answer. Answers on the free response section are scored on a scale that varies between three and ten points, depending on the length and complexity of the question. The free response scores are scaled and added together with the multiple choice scores, and then that total is scaled and distributed among the five-point AP scale.

There is no set passing score, but some colleges accept a score of 3 to place out of entry-level classes, while others require a 4 or 5. In 2015, 6.4 percent of students taking the AP Biology exam received a score of 1, 22.1 percent received a 2, 35.9 percent received a 3, 27.5 percent received a 4, and 8.2 percent received the best possible score of 5.

Evolution

Natural Selection and its Effect on Evolution

Biological evolution is the concept that a population's gene pool changes over generations. According to this concept, populations of organisms evolve, not individuals, and over time, genetic variation and mutations lead to such changes.

Darwin's Theory of Natural Selection

Charles Darwin developed a scientific model of evolution based on the idea of *natural selection*. When some individuals within a population have traits that are better suited to their environment than other individuals, those with the better-suited traits tend to survive longer and have more offspring. The survival and inheritance of these traits through many subsequent generations lead to a change in the population's gene pool. According to natural selection, traits that are more advantageous for survival and reproduction in an environment become more common in subsequent generations.

Evolutionary Fitness

Sexual selection is a type of natural selection in which individuals with certain traits are more likely to find a mate than individuals without those traits. This can occur through direct competition of one sex for a mate of the opposite sex. For example, larger males may prevent smaller males from mating by using their size advantage to keep them away from the females. Sexual selection can also occur through mate choice. This can happen when individuals of one sex are choosy about their mate of the opposite sex, often judging their potential mate based on appearance or behavior. For example, female peacocks often mate with the showiest male with large, beautiful feathers. In both types of sexual selection, individuals with some traits have better reproductive success, and the genes for those traits become more prevalent in subsequent populations.

Adaptations are Favored by Natural Selection

Adaptations are inherited characteristics that enhance survival and reproductive capabilities in specific environments. Charles Darwin's idea of natural selection explains *how* populations change—adaption explains *why*. Darwin based his concept of evolution on three observations: the unity of life, the diversity of life, and the suitability of organisms for their environments. There was unity in life based on the idea that all organisms descended from common ancestors. Then, as the descendants of the common ancestors faced changes in their environments they moved to new environments. There they adapted new features to help them in their new way of life. This concept explains the diversity of life and how organisms are matched to their environments.

An example of natural selection is found in penguins—birds that cannot fly. Over time, populations of penguins lost the ability to fly but became master swimmers. Their habitats are surrounded by water, and their food sources are in the water. Penguins that could dive for food survived better than those that could fly, and the divers produced more offspring. The gene pool changed as a result of natural selection.

Populations in Hardy-Weinberg Equilibrium

All populations have genetic diversity, but some populations aren't changing. The *gene pool* consists of all copies of every allele at every locus in every member of a population. If the allele and genotype frequencies of a population don't change between generations, the population is in a *Hardy-Weinberg (HW) equilibrium*, named for the British mathematician and German physician who came up with the concept in 1908. There are five conditions that must be met for a HW equilibrium: (1) a large population size, (2) absence of migration, (3) no net mutations, (4) random mating, and (5) absence of selection.

The HW equation calculates the frequency of phenotypes in a population that isn't evolving and is written as follows: $p^2 + 2pq + q^2 = 1$, where p is the frequency of one allele, q is the frequency of the other allele, and pq is the frequency of the alleles mixing. P and q must add up to equal 1. As in the figure below, in a given population of wildflowers, the frequency of the red flower allele (p) is 80%, and the frequency of the white flower allele (q) is 20%. Therefore, $p = 0.8$ and $q = 0.2$. In a non-evolving population, the frequency of red flowers would be $p^2 = 0.64 = 64\%$, the frequency of pink flowers as a mix of red flower and white flower alleles would be $2pq = 0.32 = 32\%$, and the frequency of white flowers would be $q^2 = 0.04 = 4\%$. If the frequency of any flower color doesn't match the calculations from the HW equation, then the population is evolving.

Parameters for Natural Selection

There are three important points to remember about natural selection. Although natural selection occurs due to an individual organism's relationship to its environment, it is a population—not individuals— that change over time. Second, natural selection only affects heritable traits that vary within a population. If all individuals within a population share an identical trait, natural selection cannot occur, and that trait will not be modified. Lastly, which traits are the favored traits is always changing. The environment is an important factor in natural selection, so if the environment changes, a trait that was previously favored may no longer be beneficial. Natural selection is a fluid process that is always at work.

Environmental Change Serve as Selective Mechanisms

The environment constantly changes, which drives selection. Although an individual's traits are determined by their *genotype*, or makeup of genes, natural selection more directly influences *phenotype*, or observable characteristics. The outward appearance or ability of individuals affects their ability to adapt to their environment and survive and reproduce. Phenotypic changes occurring in a population over time are accompanied by changes in the gene pool.

The classic example of this is the peppered moth. It was once a light-colored moth with black spots, though a few members of the species had a genetic variation resulting in a dark color. When the Industrial Revolution hit London, the air became filled with soot and turned the white trees darker in color. Birds were then able to spot and eat the light-colored moths more easily. Within just a few months, the moths with genes for darker color were better able to avoid predation. Subsequent generations had far more dark-colored moths than light ones. Once the Industrial Revolution ended and the air cleared, light-colored moths were better able to survive, and their numbers increased.

Causes of Phenotypic Variations

There are three ways in which phenotypes change over time due to natural selection: directional selection, disruptive selection, and stabilizing selection. *Directional selection* occurs when an extreme phenotypic variation is favored. This generally happens when a population's environment changes or the population migrates to a new habitat. When the Galapagos Islands suffered a drought, finches with larger beaks were able to eat the larger, tougher seeds that became abundant. Thus, finches with that phenotype survived and reproduced more often, and that trait became more prevalent in subsequent generations. *Disruptive selection* occurs when both extremes of a phenotype are favored. Finches in Cameroon have either large beaks or small beaks. The large-beaked birds are efficient at eating large, tough seeds; the small-beaked birds are adept at eating small seeds. Birds with medium-sized beaks were not adept at eating either size of seed, so selection favored the other finches. *Stabilizing selection* occurs when neither extreme phenotype is favored, and the intermediate phenotype is best suited for adapting to the population's environment. If mice live in an environment with a mix of light and dark colored rocks, mice with an intermediate fur color are favored. Neither light nor dark fur will be selected.

The Effect of Phenotypic Variations on Fitness

Geneticists have a specialized definition of fitness. They use the term to denote an organism's capacity to survive, mate, and reproduce. This ultimately equates to the probability or likelihood that the organism will be able to pass on its genetic information to the next generation. Fitness does not mean the strongest, biggest, or most dangerous individual. A more subtle combination of anatomy, physiology, biochemistry, and behavior determine genetic fitness.

Another way to understand genetic fitness is by knowing that phenotypes affect survival and the ability to successfully reproduce. Phenotypes are genetically determined and genes contributing to fitness tend to increase over time.

Thus, the "fittest" organisms survive and pass on their genetic makeup to the next generation. This is what Darwin meant by "survival of the fittest," which is the cornerstone of Darwin's theory of evolution.

Influence of Phenotype on Genotype

Natural selection provides processes that tend to increase a population's adaptive abilities. The strongest phenotypes survive, prosper, and pass on their genetic code to the next generation. The new generation of phenotypes has a fitter genotype because they have inherited more adaptive characteristics.

Thus, the fit thrive while the weak become extinct over time. This pattern, when repeated over many generations, develops strong, fit phenotypes that survive and reproduce offspring who are as fit as or fitter than their parents. Sometimes this apparently inexorable movement can be modified by drastic external conditions such as wide variations in climate. It is important to remember that all extinct species were once fit and adapted to their environments. Unforeseen circumstances always have the potential to cause chaos in the physical world.

Other Causes of Genetic Changes in Species and Populations

While the concept of natural selection focuses on changes over time in relation to the environment, there are other circumstances when change occurs randomly. Not all genetic changes relate to survival and reproduction. *Genetic drift* is the idea that the alleles of a gene can change unpredictably between generations due to chance events. If certain alleles are lost between generations, the genetic diversity of the population decreases because that genetic variation is lost forever. For example, a population of wildflowers consists of red flowers (RR and Rr) and white flowers (rr). If a large animal destroys all of the white wildflowers, the subsequent generation could be left with no (or far fewer) alleles for white flowers. Genetic drift has the greatest effect on smaller populations. Certain alleles can be over- or under-represented, even if they are not advantageous. In addition, harmful alleles can become fixed if their normal counterpart becomes extinct.

A *population bottleneck* is a type of genetic drift. This occurs when a population significantly decreases, usually due to a sudden change in the environment such as a flood or a fire. In the surviving population, certain alleles may be over- or under-represented, and others may be completely missing. Even if the surviving population returns to its original size, it will lack the genetic diversity of the original population. The *founder effect* is a special case of the bottleneck effect. It occurs when a few individuals become separated from the larger population and form their own new population. The frequency of non-dominant alleles may increase in the new smaller population, as may the frequency of inherited disorders due to a lack of dominant alleles that would keep the disorder inactive.

Evolutionary History of Species and Common Ancestry

Phylogenetics is the subfield of biology that studies the evolutionary history of a species or group of species and their relationships. Heritable traits are evaluated to make conclusions about similarities and differences between organisms. Analyzing these characteristics increases our understanding of species and population evolution.

Taxonomy

Taxonomy is the science behind the biological names of organisms. Biologists often refer to organisms by their Latin scientific names to avoid confusion with common names, such as with fish. Jellyfish, crayfish, and silverfish all have the word "fish" in their name, but belong to three different species. In the eighteenth century, Carl Linnaeus C species, called the *binomial*, which has two parts: the *genus*, which comes first, and the *specific epithet*, which comes second. Similar species are grouped into the same genus. The Linnean system is the commonly used taxonomic system today and, moving from comprehensive similarities to more general similarities, classifies organisms into their species, genus, family, order, class, phylum, kingdom, and domain. *Homo sapiens* is the Latin scientific name for humans.

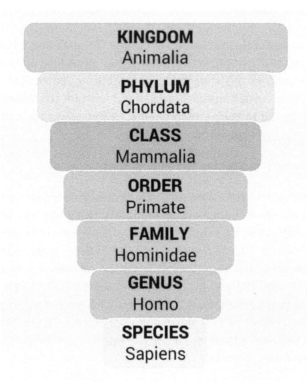

Organisms with Common Ancestry Share Features

Evolutionists propose that organisms that developed from a common ancestor often have similar characteristics that function differently. This similarity is known as *homology*. For example, humans, cats, whales, and bats all have bones arranged in the same manner from their shoulders to their digits. However, the bones form arms in humans, forelegs in cats, flippers in whales, and wings in bats, and these forelimbs are used for lifting, walking, swimming, and flying, respectively. Evolutionists look to homology, believing that the similarity of the bone structure shows a common ancestry but that the functional differences are the product of evolution.

Homologous Structures

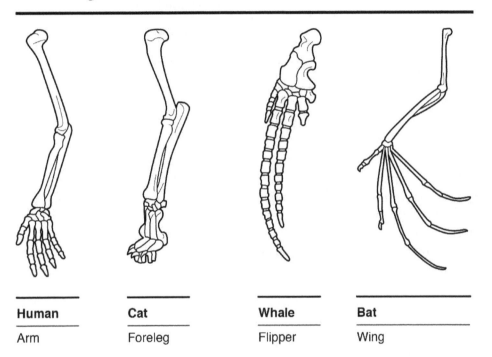

Human	Cat	Whale	Bat
Arm	Foreleg	Flipper	Wing

The Fossil Record

Fossils are the preserved remains of animals and organisms from the past, and they can elucidate the homology of both living and extinct species. Many scientists believe that fossils often provide evidence for evolution. They further propose that looking at the *fossil record* over time can help identify how quickly or slowly evolutionary changes occurred, and can also help match those changes to environmental changes that were occurring concurrently.

Species Relatedness as Illustrated with Phylogenetic Trees and Cladograms

Phylogenetic trees are branching diagrams that are used to represent the believed evolutionary history of a species. The branch points most often match the classification groups set forth by the Linnean system. Using this system helps elucidate the relationship between different groups of organisms. The diagram below is that of an empty phylogenetic tree:

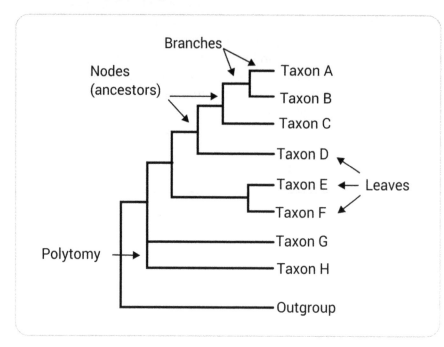

Each branch of the tree represents the divergence of two species from a common ancestor. For example, the coyote is known as *Canis latrans* and the gray wolf is known as *Canis lupus*. They share the same genus but evolved into two different species and could be represented by Taxon A and Taxon B extending from a common *Canis* branch. Groups or organisms that share a common ancestor or node in the diagram are known as *sister taxa*. A branch point leading to more than two descendent groups is known as a *polytomy*.

There are three important points to remember about phylogenetic trees. First, their purpose is to show a proposed pattern of descent, not to indicate phenotypic similarity between organisms. Second, the diagram doesn't represent time. It's believed that two taxons from the same ancestor may have become their own species at different points in history and at different rates. Lastly, taxons that are next to each other didn't come from each other. The tree diagram only indicates that it's proposed that they came from the same ancestor, which is noted at their shared node.

Cladograms

Cladistics is a method of classifying organisms based primarily on their proposed common ancestry. Using this method, species are grouped into *clades*, which include one ancestral species and all of its descendants. Some assume that species with similar traits are related. However, these similarities may appear by analogy, which means that the species were subject to similar a natural selection process but don't share a common ancestor. Cladograms help discern the difference between analogous features and homologous features. Below is an example of a cladogram:

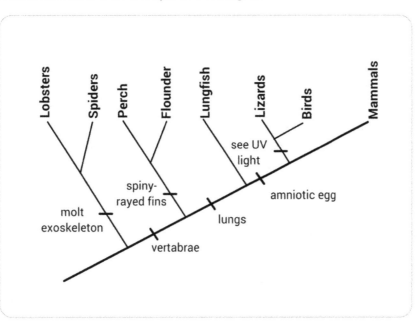

While a phylogenetic tree diagrams an organism's believed evolutionary history, a cladogram specifies the characteristics that change within the descendent groups, making it easy to see the homology of traits between related species.

Types of Speciation

Speciation is the process by which one species splits into two or more species. There are two main types of speciation that can occur: allopatric and sympatric. *Allopatric speciation* happens because of geographic separation. One population is divided into two subpopulations. If a large lake becomes divided into two smaller lakes during a drought, each lake has its own population of a species of fish that cannot intermingle with the fish in the other lake. This type of speciation can also occur when a subgroup of a population migrates to a new geographic area. When the genes of these two subpopulations are no longer mixing, new mutations can arise, and natural selection and genetic drift can take place.

In *sympatric speciation*, two or more species arise without a geographic barrier. Instead, gene flow among the population is reduced by polyploidy, sexual selection, or habitat differentiation. *Polyploidy* results when cell division during reproduction creates an extra set of chromosomes; it's more common in plants than animals. In sexual selection, organisms of one sex choose their mate based on certain traits. If there's high selection for two extreme variations of a trait, sympatric speciation may occur. For example, the females of two species of cichlid fish choose their mates based on the color of their backs—one has a blue-tinged back and the other has a red-tinged back. *Habitat differentiation* occurs when a subpopulation exploits a resource that's not used by the parent population. The North American maggot fly originally used the hawthorn tree as its habitat. A subsequent generation of this fly chose to colonize apple trees instead, creating two populations.

Both allopatric and sympatric speciation can occur quickly or slowly, and may involve a few or many genetic changes between the species.

Extinction of Species

Just as speciation creates new species, the populations of some organisms shrink and eventually become extinct. A species is considered *extinct* at the death of its very last individual. Some species cannot adapt to changes in their environment and may also find themselves weaker than new species. Speciation may generate a strong species that preys on weaker species until they are extinct. Competition between species for limited resources may also wipe out a weaker species.

Human Impact on Ecosystems

Humans are responsible for many harmful changes to the environment. Deforestation and urbanization cause species to lose the environment to which they are adapted. Overfishing and overhunting greatly decrease the population of many species. Pollution also causes environmental destruction to which species cannot adapt.

Reproductive Isolation Can Impact Speciation

One important distinguishing factor in the formation of two species is their reproductive isolation. Species are characterized by their members' ability to breed and produce viable offspring. When speciation occurs and new species are formed, there must be a biological barrier that prevents the two species from producing viable offspring.

Following speciation, there are two types of *reproductive barriers* that keep the two populations from mating with each other. These are classified as either prezygotic barriers or postzygotic barriers. *Prezygotic barriers* prevent fertilization via habitat isolation, temporal isolation, and behavioral isolation. Through habitat isolation, two species may inhabit the same area but don't often encounter each other. *Temporal isolation* is when species breed at different times of the day, during different seasons, or during different years, so their mating patterns never coincide. *Behavioral isolation* refers to mating rituals that prevent an organism from recognizing a different species as potential mate.

Other prezygotic barriers block fertilization after a mating attempt. *Mechanical isolation* occurs when anatomical differences prevent fertilization. *Gametic isolation* occurs when the gametes of two species are incompatible.

Postzygotic barriers contribute to reproductive isolation after fertilization. The mixed-species offspring are called *hybrids*. Once the hybrid zygote is formed, it may have reduced viability because of differing numbers of chromosomes between the parents. When meiosis takes place, normal gamete cells aren't produced in the offspring. These hybrids may not survive, or if they do, they are very frail. If a hybrid survives and becomes a healthy animal, it may have reduced fertility. For example, when a male donkey mates with a female horse, they produce a mule. Mules are healthy but sterile. When a female donkey mates with a male horse, they produce a hinny, which is also sterile. *Hybrid breakdown* is another postzygotic barrier. In some cases, hybrids that are produced are actually viable and fertile for the first generation. However, when the hybrids mate with each other or one of the parent species, the subsequent generations are weak or sterile.

Evolution is Continuous

Although natural selection supports the survival and reproduction of the individuals best suited for their environment, their variation always remains within a population. This idea is known as balancing selection. Balancing selection maintains the variation in a population through the heterozygote advantage and frequency-dependent selection. Because the environment is always changing, the characteristics necessary to increase or maintain survival of a species may also change.

Heterozygote Advantage

The heterozygote advantage refers to when individuals that are heterozygous for a particular trait are more fit for their environment than individuals that are homozygous for either variation of the trait. An example is the gene that codes for the β-subunit of hemoglobin, the oxygen-carrying protein of red blood cells. The homozygous genotype causes sickle-cell disease, which distorts the shape of red blood cells so they block the flow of blood. The heterozygous genotype for this subunit of hemoglobin protects humans against the most severe effects of malaria because the body can quickly destroy the distorted red blood cells and get rid of the parasites carrying malaria. The human population is now predominantly heterozygous for this gene.

Frequency-Dependent Selection

In frequency-dependent selection, the occurrence of a phenotype depends on how common it is in the population. If a phenotype becomes too common, it may no longer benefit the survival and reproduction of the species. For example, there are scale-eating fish in Lake Tanganyika in Africa that can be either right-mouthed or left-mouthed. Left-mouthed fish attack their prey's right flank, and right-mouthed fish attack their prey's left flank. The prey species learn to protect themselves from the type of scale-eating fish most common in the lake. Therefore, the dominance of the left- and right-mouthed fish in the lake changes over time and keeps the species from becoming extinct due to lack of prey. Balancing selection keeps both alleles of the gene present in the scale-eating fish population, and is propose by some to play a large role in their evolution.

Hypotheses About the Origin of Life of Earth

Abiogenesis is the term used to refer to the theoretical process by which life developed from non-living matter, such as organic compounds. Many scientists estimate that the Earth is approximately 4.5 billion years old and believe that the first living organisms developed between 3.8 and 4.1 billion years ago. There are several theories about how life formed on Earth.

One theory on the origin of life involves the creation of organic compounds from a combination of minerals from the sea and ideal atmospheric conditions. In the deep sea, hydrothermal vents release minerals from the Earth's interior along with hot water. Water from the alkaline vents on the sea floor has a high pH and provides a stable environment for organic compounds. In addition, theorists propose that the Earth's atmosphere had reducing qualities that could produce organic compounds from simpler molecules. They further believe that the warm atmosphere above volcanoes was ideal for this synthesis of organic compounds. The *primordial soup* theory proposes that larger and more complex compounds were made over time from these original organic compounds, eventually forming living organisms.

Panspermia is the concept that life came to Earth from other areas of the universe. It hypothesizes that meteoroids, asteroids, and other small objects from space landed on Earth and transferred microorganisms to the Earth's surface. This theory proposes that there were seeds of life everywhere in the universe, and when these seeds were brought to Earth, the conditions were ideal for living organisms to develop and flourish.

Protocells are small, round groups of lipids hypothesized to be responsible for the origin of life. *Vesicles* are fluid-filled compartments enclosed by a membrane-like structure. They form spontaneously when lipids, such as protocells, are added to water and have important features for the creation of living organisms, such as dividing on their own to form new vesicles and absorbing material around them. When vesicles encounter lipids, the lipids form a bilayer around the vesicle, similar to a plasma membrane of a cell. According to scientists, these cells were able to encapsulate minerals and organic molecules around them. Some of the clay that covered the primordial Earth was believed to be covered in RNA, which the vesicles could encapsulate. These simple behaviors are believed to have given rise to more complex behaviors, such as simple cell metabolism, and began resembling true cells, now known as protocells. As the cells interacted with each other, theorist propose that larger living organisms were created.

Evolutionists propose that life on Earth began with RNA. Although RNA is a genetic material, it can also be an enzyme-like catalyst, known as a *ribozyme*. Ribozymes can catalyze chemical reactions and self-replicate to make complementary copies of short pieces of RNA. According to the theory of evolution, vesicles that carried RNA could then divide and have replicated RNA in its daughter cell, increasing the amount of genetic material in the environment. These daughter cells would have been protocells. Evolutionists believe that the RNA inside of them were likely used as templates to then create more stable DNA strands. It's proposed that then from the formation of DNA, the origin of life and more complex living organisms began.

In 2015, scientists Sergei Maslov and Alexei Tkachenko expanded on this theory. They believe that the self-replicating model was cyclical and went through different phases during the day and night. During the day, the polymers would float freely, while at night, the polymer chains would join together to form longer polymers using a template, a process called *template-assisted ligation*. They believe that although the chains could join without a template, the use of a template is more efficient and reproducible for preserving the original sequences. According to these two scientists, these phases occurred at different times due to changes in the environment, such as with temperature, pH, and salinity. These factors then regulated whether the polymers would come together or float apart.

Scientific Evidence of Origin Theories
Over the years, many scientific experiments have attempted to prove that different theories about the origin of life are true. In 1953, the Miller-Urey chemical experiment simulated what they believe to be the atmospheric conditions of early Earth. It was believed that the atmosphere contained water, methane, ammonia, and hydrogen. Scientists Stanley Miller and Harold Urey showed that an electrical spark, like a bolt of lightning, helped catalyze the creation of complex organic compounds from simpler ones. They hypothesized that the complex molecules would then react with each other and the simple compounds to form even more compounds, such as formaldehyde, hydrogen cyanide, glycine, and sugars to produce life.

In support of self-replicating RNA, several scientists have tried to create the shortest RNA chain possible that can replicate itself. In the 1960s, Sol Spiegelman created a short RNA chain consisting of 218 bases that was able to replicate itself with an enzyme from a 4500 base bacterial RNA. In 1997, Manfred Eigen was able to further degrade a large RNA chain to only approximately 50 bases, which was the minimum length needed to bind a replication enzyme. Similarly, researchers at the J. Craig Venter Institute have used engineering techniques to try to create prokaryotic cells with as few genes as possible to figure out the minimal requirements for life. In 1995, they started with a microbe with the smallest genome known to humans with 470 genes and were able to take away one gene at a time to leave only 375 essential genes.

Practice Questions

1. Charles Darwin's theory of evolution is based on what type of selection?
   a. Natural selection
   b. Sexual selection
   c. Disruptive selection
   d. Stabilizing selection
   e. Allopatric selection

2. Which of the following phrases best fits into the following sentence?
Natural selection makes individuals _____ their environments.
   a. less fit to
   b. less affected by
   c. grow faster in
   d. eat more in
   e. more adaptable to

3. Which type of selection describes the finches of the Galapagos Islands developing larger beaks so they are able to eat the larger, tougher seeds that became abundant after a drought?
   a. Sexual
   b. Stabilizing
   c. Directional
   d. Disruptive
   e. Adaptable

4. When mice develop an intermediate fur color instead of light or dark fur, what type of selection is occurring?
   a. Disruptive
   b. Stabilizing
   c. Directional
   d. Sexual
   e. Adaptable

5. The founder effect occurs when which of the following occur?
   a. A new species suddenly fills an open niche.
   b. A new species is developed.
   c. Individuals develop an extreme phenotype through natural selection.
   d. There is a sudden change in environmental conditions.
   e. A few individuals become separated from the larger population and form a new population.

6. Which circumstance of random change leaves a population less diverse than its original composition?
   a. A geographical barrier
   b. Founder effect
   c. A strong wind
   d. Bottleneck effect
   e. Speciation

7. What do evolutionary theorist believe the Hardy-Weinberg equation tell us?
   a. Whether a population is evolving
   b. The type of natural selection occurring
   c. If genetic drift is altering a population
   d. The size of the population
   e. The expected rate of population growth.

8. What is an adaptation?
   a. The original traits found in a common ancestor
   b. Changes that occur in the environment
   c. When one species begins behaving like another species
   d. An inherited characteristic that enhances survival and reproduction
   e. Changes that occur in an individual during the aging process

9. What is the broadest, or LEAST specialized, classification of the Linnean taxonomic system?
   a. Species
   b. Family
   c. Kingdom
   d. Phylum
   e. Class

10. According to evolutionary theory, what are vestigial structures?
    a. Anatomical structures that stick out of the body
    b. Structures found only in the foot
    c. Structures found only in the hand
    d. Anatomical structures that are still present but no longer have a function
    e. Structures that give a species a phenotypical advantage in their environment

11. Which taxonomic system is commonly used to describe the hierarchy of similar organisms today?
    a. Aristotle system
    b. Linnean system
    c. Cesalpino system
    d. Darwin system
    e. The Hardy-Weinberg system

12. What is the Latin specific name for humans?
    a. Homo sapiens
    b. Homo erectus
    c. Felis catus
    d. Canis familiaris
    e. Homo habilis

13. What do phylogenetic trees tell us about a species?
    a. The genetic contribution of each allele that an offspring inherits
    b. The believed size of the population
    c. How many alleles exist for a specific trait of the species
    d. Their eye color
    e. Their proposed evolutionary history

14. How do cladograms differ from phylogenetic trees?
    a. They include pictures of the species
    b. They specify common species names instead of Latin names
    c. They are not branching diagrams
    d. They specify the characteristics that changed in descendent groups, creating new species
    e. They show the specific alleles that each parent had, and the potential crosses that could result

15. What is speciation?
    a. The process by which a species becomes extinct
    b. The relationship between two species as described by a phylogenetic tree
    c. The act of drawing a phylogenetic tree
    d. The process by which a predator overtakes its prey
    e. The process by which one species splits into two or more species

16. In allopatric speciation, what causes a species to split in two?
    a. A geographic barrier
    b. Natural selection
    c. A change in environment
    d. Interspecies mating
    e. Competition

17. Sexual selection stops gene flow and causes what type of speciation?
    a. Allopatric
    b. Natural
    c. Sympatric
    d. Sexual
    e. Prezygotic

18. In addition to pollution and deforestation, what other human activities can cause extinction?
    a. Urbanization and overfishing
    b. Urbanization and competition
    c. Predation and urbanization
    d. Overfishing and competition
    e. Predation and competition

19. When is a species considered extinct?
    a. At the death of the last individual
    b. When there are fewer than 10 individuals remaining
    c. When the species has not had any individuals alive for 5 years
    d. When exactly 2 individuals remain
    e. When the death rate exceeds the birth rate for at least 5 years

20. Which is an example of a prezygotic reproductive barrier?
    a. Changing environments
    b. Three species inhabiting the same area
    c. Large feathers
    d. Habitat isolation
    e. Temperature changes

21. What type of barrier leads to reproductive isolation after two species mate and produce a hybrid offspring?
   a. Postzygotic barrier
   b. Habitat isolation
   c. Temporal isolation
   d. Behavioral isolation
   e. Mechanical isolation

22. Which concept maintains the variation in a population even as natural selection occurs?
   a. Genetic drift
   b. Sexual selection
   c. Postzygotic barriers
   d. Balancing selection
   e. Bottleneck

23. Which idea about the origin of life hypothesizes that microorganisms were transferred to Earth from other objects in the solar system?
   a. Primordial soup
   b. Self-replicating RNA
   c. Endosymbiosis
   d. Protocells
   e. Panspermia

24. Protocells are an essential vesicle for replicating what material essential to a hypothesis about the origin of life on Earth?
   a. Clay
   b. RNA
   c. Chromosomes
   d. Carbohydrates
   e. Chloroplasts

25. What did the Miller-Urey experiment simulate to test the creation of complex compounds from simpler ones that are believed to have existed on the surface of early Earth?
   a. The atmospheric conditions
   b. The Earth's interior temperature
   c. The oceanic environment
   d. The number of leaves on each tree
   e. The nature of algae in deep sea trenches

26. What is a driving force behind why speciation can occur?
   a. Geographic separation
   b. Seasons
   c. Daylight
   d. A virus
   e. Warm temperatures

Answer Explanations

1. A: Charles Darwin founded the theory of natural selection. He believed that stronger individuals would continue to thrive while weaker individuals would die off.

2. E: Natural selection is the idea that individuals within a population can survive longer and with higher reproduction rates based on specific traits that they've inherited that make them better matched to their environment.

3. C: Natural selection acts on the phenotypes of individuals. Directional selection occurs when one extreme of the phenotypic variations is favored. This generally happens when a population's environment changes or the population migrates to a new habitat.

4. B: Stabilizing selection occurs when neither extreme phenotype is favored and the intermediate phenotype is most suitable for adapting to the population's environment. If mice live in an environment with a mix of light- and dark-colored rocks, their fur will be an intermediate color. Neither light nor dark fur will be selected.

5. A: The founder effect occurs when a few individuals from a population become separated from the larger population and form their own new population. This may occur when a storm blows a few individuals to a new island or habitat. The frequency of non-dominant alleles may increase in the new smaller population, as well as the frequency of inherited disorders due to a lack of dominant alleles that would keep the disorder inactive.

6. D: The bottleneck effect occurs when there's a sudden change in the environment. A flood or a fire could drastically decrease the size of a population. In the surviving population, certain alleles may be over- or under-represented, and others may be completely gone. Even if the surviving population reaches back to its original size over time, it will lack the genetic diversity of the original population.

7. A: All populations have genetic diversity, but according to evolutionists, that doesn't guarantee the population is evolving. In order to assess whether a population is evolving, scientists use a mathematical equation to calculate the phenotypes of a non-evolving population. The results of that equation can be compared to the actual phenotypes seen in the population. If the allele and genotype frequencies of a population aren't changing between generations, the population is described as being in a Hardy-Weinberg (HW) equilibrium.

8. D: Charles Darwin based the idea of adaptation around his original concept of natural selection. He believed that evolution occurred based on three observations: the unity of life, the diversity of life, and the suitability of organisms to their environments. There was unity in life based on the idea that all organisms descended from a common ancestor. Then, as the descendants of common ancestors faced changes in their environments or moved to new environments, they began adapting new features to help them. This concept explained the diversity of life and how organisms were matched to their environments. Natural selection helps to improve the fit between organisms and their environments by increasing the frequency of features that enhance survival and reproduction.

9. C: In the Linnean system, organisms are classified as follows, moving from comprehensive and specific similarities to fewer and more general similarities: species, genus, family, order, class, phylum, kingdom, and domain. A popular mnemonic device to remember the Linnean system is "Dear King Philip Came Over for Good Soup."

10. D: Evolutionists propose that structures present in descendent species that no longer have a function are known as vestigial structures. For example, some snakes still have pelvis and leg bones that descended from ancestors that walked.

11. B: The Linnean system is the commonly used taxonomic system today. It classifies species based on their similarities and moves from comprehensive to more general similarities. The system is based on the following order: species, genus, family, order, class, phylum, and kingdom.

12. A: Homo is the human genus. Sapiens are the only remaining species in the homo genus.

13. A: Phylogenetic trees are used to illustrate the believed evolutionary history of a species. They are branching diagrams, and the branch points most often match the classification groups set forth by the Linnean system. Using this system helps elucidate the relationship between different groups of organisms.

14. D: Cladograms help discern the difference between analogous features and homologous features. While a phylogenetic tree diagrams the believed evolutionary history of an organism, a cladogram specifies the characteristics that change within descendent groups, making it is easy to see the homology of certain traits between related species.

15. E: Speciation is the process by which one species splits into two or more species.

16. A: Allopatric speciation happens because of geographic separation. One population is divided into two subpopulations. If a drought occurs and a large lake divides into two smaller lakes, each lake is left with its own population that cannot intermingle with the population of the other lake. This type of speciation can also occur when a subgroup of a population migrates to a new geographic area.

17. C: In sympatric speciation, two or more species arise without a geographic barrier. In sexual selection, organisms choose their mate based on certain traits. If there is high selection for two extreme variations of a trait, sympatric speciation may occur. The females of two distinct species of cichlid fish choose their male mates based on the color of their backs—one has a blue-tinged back and the other has a red-tinged back.

18. A: Humans are responsible for many harmful changes to the environment. Deforestation and urbanization cause species to lose their adapted environment all together. Overfishing and overhunting greatly decrease the number of individuals in a species. Pollution also cause destructive changes to the environment to which some species may be unable to adapt.

19. A: A species is considered extinct at the death of its very last individual. If any number of individuals are still alive, the species may be considered endangered, but not extinct.

20. D: Prezygotic barriers prevent fertilization. These include habitat isolation, temporal isolation, and behavioral isolation. Two species may inhabit the same area but don't encounter often each other, which is habitat isolation.

21. A: Postzygotic barriers contribute to reproductive isolation after fertilization. The mixed species offspring are called hybrids. Once the hybrid zygote is formed, it may have reduced viability. These hybrids may not survive, and if they do, they are very frail. If a hybrid survives, it may have reduced fertility. In some cases, hybrids are actually viable and fertile for the first generation. However, when the hybrids mate with each other or with one of the parent species, subsequent generations are weak or sterile.

22. D: Although the idea of natural selection supports the survival and reproduction of the individuals best suited for their environment, there's always variation within a population. This idea is known as balancing selection, which maintains the variation in a population through the heterozygote advantage and frequency-dependent selection. As the environment is always changing, the characteristics necessary to increase or maintain survival and reproduction of a species may also change.

23. E: Panspermia is the idea that life came to Earth from other areas of the universe. It hypothesizes that meteoroids, asteroids, and other small objects from the solar system landed on Earth and transferred microorganisms to the Earth's surface. This theory proposes that there were seeds of life everywhere in the universe, and when these seeds were brought to Earth, the conditions were ideal for living organisms to develop and flourish.

24. B: Protocells are small, round groups of lipids that are hypothesized to be responsible for the origin of life. They are proposed to have been formed by vesicles, which are fluid-filled compartments enclosed by a membrane-like structure. Vesicles form spontaneously when lipids, such as protocsells, are added to water. Some of the clay that covered the Earth is believed to contain RNA, which the vesicles could encapsulate. Vesicles that carried RNA could then divide and have replicated RNA in its daughter cell, increasing the amount of genetic material in the environment. The RNA inside of them may have been used as templates to create more stable DNA strands. Then, according to evolutionary theory, after the formation of DNA, the origin of life and more complex living organisms began.

25. A: In 1953, the Miller-Urey experiment attempted to simulate the atmospheric conditions of early Earth. It was believed that the atmosphere contained water, methane, ammonia, and hydrogen. Scientists Stanley Miller and Harold Urey believed an electrical spark, such as a bolt of lightning, helped catalyze the creation of complex organic compounds from simpler ones. The complex molecules would then react with each other and the simple compounds to form even more compounds, such as formaldehyde, hydrogen cyanide, glycine, and sugars.

26. A: Speciation is the method by which one species splits into two or more species. In allopatric speciation, one population is divided into two subpopulations. If a drought occurs and a large lake becomes divided into two smaller lakes, each lake is left with its own population that cannot intermingle with the population of the other lake. When the genes of these two subpopulations are no longer mixing with each other, new mutations can arise and natural selection can take place.

Biological Systems and Use of Energy

Life is Highly Organized

Organisms are highly complex entities that seemingly defy the second law of thermodynamics, which states that the universe must increase in entropy (disorder) as time goes on. If randomness and disorder are favored, it is miraculous that complicated organisms, such as humans, could ever exist, since the processes that maintain life are very orderly.

Input of Free Energy into the System Drives the Processes of Life

Life processes require energy. Energy is defined as the ability to do work or move matter against external forces. Free energy is the energy within a physical system that is used to do work. There are two basic forms of energy: potential and kinetic. *Potential energy* is the energy stored within an object that has the capacity to do work, but hasn't yet. For example, consider two equally strong people pulling a rope on opposite ends as hard as they can. While they are in this position, nothing is happening. However, energy stored as tension in the rope has the *potential* to do work. If one person suddenly let go of the rope, the person on the other end would fall backward. Once this person falls, the potential energy is now converted into *kinetic energy* – the energy of movement. Potential energy is the most important type of energy in biochemical reactions, since high amounts of potential energy stored in the atomic bonds of molecules can be metabolized to release kinetic energy, which can be harnessed to do many different kinds of work.

Entropy and Maintaining Homeostasis

In order for organisms to survive against this universal tendency for chaos, they must use energy to work against entropy. They do this via biochemical processes that maintain an internal order, called homeostasis. Homeostasis is the physiological processes within a system that regulate a stable internal equilibrium such as body temperature, blood pH, and fluid balance.

Organisms maintain homeostasis by using free energy and matter, usually in the form of atoms and molecules. Atoms are the smallest whole units of matter, and molecules are atoms connected to each other via chemical bonds. Everything is made up of atoms, even the most complicated structures; their complexity is only due to the variety and quantity of different molecular arrangements. A famous experiment by Stanley Miller and Harold Urey simulated early Earth's atmospheric conditions and witnessed the synthesis of a string of molecules known as amino acids. Amino acids are the blueprints of proteins — the building blocks of life. This famous experiment provided a glimpse of the original atomic arrangements that facilitated the origin of life. Before proteins, random particles would collide, due to chance, only if given enough time. After proteins arrived, life arose, likely because proteins enable organisms to use available energy and matter to make complicated structures, from cells to whole organisms. Enzymes (proteins that act as catalysts) help reactants find each other by providing a docking station, increasing probabilities of atomic collision and bonding, and therefore, facilitate the specific biochemical reactions that make life possible.

Here's an illustration of that:

The Lock and Key Mechanism

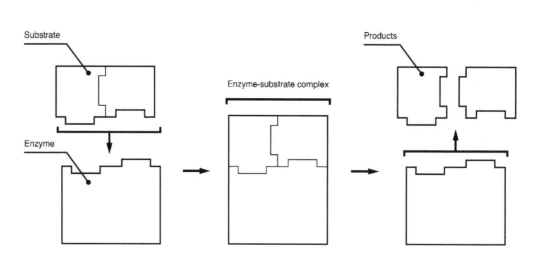

Proteins combine to make a myriad of different structures. Think of proteins like people. A pile of bricks can't do anything on its own. Buildings can only form when there are workers to move the bricks into place. Similarly, proteins turn matter into the complicated structures that form organisms.

Proteins in the form of enzymes are especially important for biochemical reactions because they lower the activation energy — the minimum energy required for a chemical reaction to happen — as illustrated in the below graph:

Using available free energy, reactions occur within an organism that create organization and decrease entropy. The below graphs illustrate how free energy is used in two basic biochemical reactions:

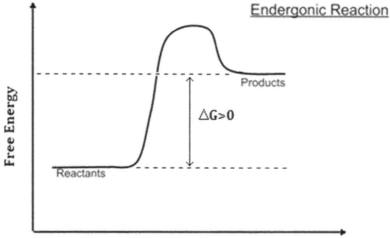

The image above illustrates a reaction that decreases entropy (or increases order), called an *endergonic reaction*. Endergonic, or anabolic, reactions occur when reactants *absorb* energy from the surroundings so that the products hold more energy than the reactants. Anabolic reactions enable the organism to make bigger things (polymers) from smaller things (monomers), such as forming new cells, or making proteins from amino acids.

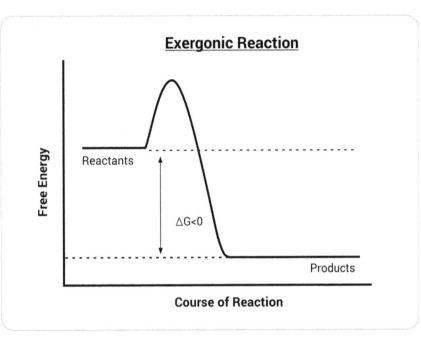

Conversely, the image above illustrates an exergonic, or catabolic, reaction that *releases* energy. In this reaction, the products hold less energy than the reactants, and entropy (or disorder) is increased. Catabolic reactions enable organisms to break down bigger things (polymers) into smaller things (monomers); for example, breaking down proteins into their respective amino acids.

Maintain Order and Cellular Processes, Energy Input Must Exceed Energy Lost

As long as the energy released (exergonic) exceeds the energy absorbed (endergonic) in an organism, it will continue to function and sustain life. If there is not energy available from exergonic reactions to drive endergonic reactions, the organism will die, since energy is required to perform all metabolic functions. Metabolism is the set of processes carried out by an organism that permits the exchange of energy between itself and the environment, enabling it to change and grow internally.

In other words: metabolism = catabolism + anabolism

- If catabolism is greater than anabolism, i.e., energy released is greater than energy consumed, then an organism can live.

- If catabolism is less than anabolism, i.e., energy released is less than energy consumed, then an organism will die.

Organisms must consume energy, in the form of food or light, to perform anabolic reactions. If they are autotrophs, such as plants, they produce their own food. If they are heterotrophs, such as animals, they absorb the energy provided by food, most commonly sugars. The most common form of sugar used for energy is the simple sugar glucose, C6H12O6, which contains extremely high amounts of potential energy stored within its atomic bonds.

Cellular Respiration

Cellular respiration is the catabolic process of breaking down the bonds in glucose and releasing its potential energy in the form of ATP, or adenosine triphosphate. ATP harnesses small amounts of energy and uses it for processes in cellular metabolism. Each glucose molecule can produce about 32 ATP molecules. Breaking glucose and storing its energy in smaller molecules enables the cells to distribute energy across many metabolic reactions instead of just one.

ATP holds energy in the bonds between its phosphates. It also cycles back and forth between harnessing and distributing energy by forming and breaking a phosphate bond, as shown in the figure below.

The ATP - ADP Cycle

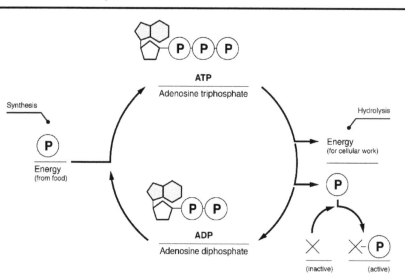

Exergonic Reactions Can Drive Endergonic Reactions

Cells balance their energy resources by using the energy from exergonic reactions to drive endergonic reactions forward, a process called energy coupling. Adenosine triphosphate, or ATP, is a molecule that is an immediate source of energy for cellular work. When it is broken down, it releases energy used in endergonic reactions and anabolic pathways. ATP breaks down into adenosine diphosphate, or ADP, and a separate phosphate group, releasing energy in an exergonic reaction. As ATP is used up by reactions, it is also regenerated by having a new phosphate group added onto the ADP products within the cell in an endergonic reaction.

Even within a single process, there are energy investments and divestments in many necessary reactions along the way. ATP is necessary for all of these reactions. It can be made in two different ways: substrate-level phosphorylation and oxidative phosphorylation.

Substrate-level phosphorylation occurs when a phosphoryl group (PO3) is donated to an ADP (adenosine diphosphate) molecule from a substrate by using enzymes, thereby creating ATP, as shown below in the image on the left. In oxidative phosphorylation, as illustrated on the right, a free phosphate joins ADP via the energy provided by chemiosmosis and the mechanical movement of the ATP synthase (the enzyme that creates ATP).

Obtaining energy at the cellular level and understanding how energy investment is used to create organization in living things, is important in order to understand energy exchange between all of the cells of an organism and even between an organism and its environment.

Energy-Related Pathways in Biological Systems

Krebs Cycle

In aerobically respiring eukaryotic cells, the pyruvate molecules then enter the mitochondrion. Pyruvate is oxidized and converted into a compound called acetyl-CoA. This molecule enters the citric acid cycle to begin the process of aerobic respiration.

The citric acid cycle has eight steps. Remember that glycolysis produces two pyruvate molecules from each glucose molecule. Each pyruvate molecule oxidizes into a single acetyl-CoA molecule, which then enters the citric acid cycle. Therefore, two citric acid cycles can be completed and twice the number of ATP molecules are generated per glucose molecule.

Eight Steps of the Citric Acid Cycle
Step 1: Acetyl-CoA adds a two-carbon acetyl group to an oxaloacetate molecule and produces one citrate molecule.

Step 2: Citrate is converted to its isomer isocitrate by removing one water molecule and adding a new water molecule in a different configuration.

Step 3: Isocitrate is oxidized and converted to α-ketoglutarate. A carbon dioxide (CO_2) molecule is released and one NAD+ molecule is converted to NADH.

Step 4: α-Ketoglutarate is converted to succinyl-CoA. Another carbon dioxide molecule is released and another NAD+ molecule is converted to NADH.

Step 5: Succinyl-CoA becomes succinate by the addition of a phosphate group to the cycle. The oxygen molecule of the phosphate group attaches to the succinyl-CoA molecule and the CoA group is released. The rest of the phosphate group transfers to a guanosine diphosphate (GDP) molecule, converting it to guanosine triphosphate (GTP). GTP acts similarly to ATP and can actually be used to generate an ATP molecule at this step.

Step 6: Succinate is converted to fumarate by losing two hydrogen atoms. The hydrogen atoms join a flavin adenine dinucleotide (FAD) molecule, converting it to $FADH_2$, which is a hydroquinone form.

Step 7: A water molecule is added to the cycle and converts fumarate to malate.

Step 8: Malate is oxidized and converted to oxaloacetate. One lost hydrogen atom is added to a NAD molecule to create NADH. The oxaloacetate generated here then enters back into step one of the cycle.

At the end of glycolysis and the citric acid cycles, four ATP molecules have been generated. The NADH and $FADH_2$ molecules are used as energy to drive the next step of oxidative phosphorylation.

<u>The Calvin Cycle</u>

There are three phases in the Calvin cycle: carbon fixation, reduction, and regeneration of the CO_2 acceptor. Carbon fixation is when the first carbon molecule is introduced into the cycle, when CO_2 from the air is absorbed by the chloroplast. Each CO_2 molecule enters the cycle and attaches to ribulose bisphosphate (RuBP), a five-carbon sugar. The enzyme RuBP carboxylase-oxygenase, also known as rubisco, catalyzes this reaction. Next, two three-carbon 3-phosphoglycerate sugar molecules are formed immediately from the splitting of the six-carbon sugar.

Next, during the reduction phase, an ATP molecule is reduced to ADP and the phosphate group attaches to 3-phosphoglycerate, forming 1,3-bisphosphoglycerate. An NADPH molecule then donates two high-energy electrons to the newly formed 1,3-bisphosphate, causing it to lose the phosphate group and become glyceraldehyde 3-phosphate (G3P), which is a high-energy sugar molecule. At this point in the cycle, one G3P molecule exits the cycle and is used by the plant. However, to regenerate RuBP molecules, which are the CO_2 acceptors in the cycle, five G3P molecules continue in the cycle. It takes three turns of the cycle and three CO_2 molecules entering the cycle to form one G3P molecule.

In the final phase of the Calvin cycle, three RuBP molecules are formed from the rearrangement of the carbon skeletons of five G3P molecules. It is a complex process that involves the reduction of three ATP molecules. At the end of the process, RuBP molecules are again ready to enter the first phase and accept CO_2 molecules.

Although the Calvin cycle is not dependent on light energy, both steps of photosynthesis usually occur during daylight, as the Calvin cycle is dependent upon the ATP and NADPH produced by the light reactions, because that energy can be invested into bonds to create high-energy sugars. The Calvin cycle invests nine ATP molecules and six NADPH molecules into every one molecule of G3P that it produces. The G3P that is produced can be used as the starting material to build larger organic compounds, such as glucose.

Free Energy is Used for Life Processes

Regardless of size, energy provided by cellular respiration is used to maintain homeostasis on a larger scale. Energy is used to run all of an organism's processes, such as digestion, brain processing, and circulation. These processes alone, not to mention all of the other bodily functions, require quite a bit of energy to sustain. Because of this, organisms have ways of conserving energy, as well as energy reserves stored in lipids, or fats, that can be used if there is insufficient energy in times of stress, but these reserves are not infinite. When energy reserves are eventually exhausted, an organism will not have enough energy to sustain life.

Accomplishment of Regulation with Feedback Loops

Feedback loops are complex regulators that keep a balance between energy expenditure and energy conservation. One example of a feedback loop includes the complex mechanisms that maintain a human's optimal body temperature. This intricate feedback loop is controlled by the hypothalamus – the link between the nervous system and endocrine system in the brain. Upon receiving a neural stimulus, the hypothalamus stimulates or inhibits the pituitary gland based on nervous system input, which is how it regulates hormone levels that have specific physiological effects. When body temperature is too low, the body is prompted to shiver, which releases energy from the muscles and warms the body. When body temperature is too high, the body initiates sweating, which regulates temperature through a process known as evaporative cooling. Evaporative cooling occurs because the water droplets heated by the body and released by sweat glands are the ones with the highest kinetic energy. The high kinetic energy results in the evaporation of the hottest water molecules, which leaves the cooler ones behind. Other endotherms, or organisms that regulate their temperature internally, have different mechanisms to cool down. Dogs, for example, only have sweat glands in their paws and nose, so they pant to use evaporative cooling by drawing moisture from their lungs.

Endothermy Versus Ectothermy for Temperature Regulation

Endotherms, such as humans, regulate body temperature internally. Conversely, ectotherms, such as reptiles, have no internal mechanism for temperature regulation. They bask in the sun or hide in the shade to regulate their body temperature. They have a significantly slower metabolism and require far less food than endotherms because they do not require advanced mechanisms to regulate temperature.

Reproduction Requires Energy

Sustaining life through energy conservation would be futile if there were no life to begin with. Every organism on the planet exists because of reproduction. In every known species on Earth, there is an inherent drive to reproduce, which requires a great deal of energy. If organisms do not expend energy procreating and ensuring the survival of their offspring, they are in danger of not passing on their genes. However, an organism cannot reproduce if it does not survive itself, so it must prioritize survival over reproduction. If survival seems unlikely, an animal will expend energy to fight or flee before it will reproduce.

For example, many bird species have elaborate courtship rituals that take a significant amount of energy in the form of dancing, squawking, and jumping around. If a male bird is about to seduce a female, but a predator approaches, he will use any available energy to flee. Energy at that time is best used to preserve his life. However, if no predator is present, he will use this energy to attempt to reproduce.

Hibernation as a Reproduction and Energy Conservation Strategy

Hibernating animals are another example, since they are generally out of reproductive commission in the winter. Their internal temperature drops to around freezing, and their metabolic rate decreases as much as 20 times lower than during arousal. This is to their benefit. They spend their limited energy stores solely on personal survival and abandon any actions toward procreation and nurturing. Similar to hibernation, estivation is the opposite trend with the same effect. This is when animals retreat in the summer to avoid the heat and water scarcity, and in doing so, they disregard any reproductive responsibilities.

The Effect of Body Size on Metabolic Rate

The metabolic rate is the amount of energy used by an organism to perform reactions in a given time period. One might assume that the larger the organism, the higher the metabolic rate, but actually, the size of the organism is inversely proportional to its metabolic rate. Small organisms have a greater surface area to volume (SA:V) ratio than larger organisms. The surface area is the total area of the outermost layer of an object, and is the area through which organisms lose heat. To illustrate this point, think of a sphere. The volume of a sphere ($4/3\pi r^3$) is much greater than its surface area ($4\pi r^2$), and increases at a much faster relative rate. With a huge sphere, there's a lot more matter inside the sphere than on the outermost layer of it. However, with a small sphere, there isn't as much matter on the inside compared to its surface layer. Therefore, the larger organisms become, the smaller the SA:V ratio. Smaller organisms have a larger SA:V ratio and hence, lose heat much faster than larger organisms. The greater metabolic rate of smaller organisms is required in order to maintain a constant body temperature.

Ecosystems Are Affected By Changes In Free Energy

Energy balance is not just a theme observed in cellular processes, organism homeostasis, and reproductive strategies; it is also a theme in ecosystems. For example, sunlight exposure variations in an ecosystem affect the ecosystem's organisms, since differences in light exposure will affect plants' glucose production. If there are fewer plants to capture the sun's energy, there is less energy available to consumers, and the carrying capacity of that ecosystem will fall. When the carrying capacity of the ecosystem falls, only the most energy-savvy organisms will survive. In fact, many organisms engage in mutualistic symbiotic relationships that aid in survival. Birds and bees and flowering plants, for example, help each other. The birds and bees receive sugary sweet nectar in return for helping the plants with pollination.

Capturing and Storing Free Energy

Energy is captured and stored within the building blocks of organisms called cells. There are two distinct types of cells: *prokaryotic* (found in bacteria) and *eukaryotic* (found in plants and animals). A prokaryotic cell is a single-celled organism with no nucleus, whereas a eukaryotic cell contains a nucleus and can be a whole organism or exist in multicellular organisms. A prokaryotic cell has few organelles (specialized structures within cells that perform important functions, such as chloroplasts and mitochondria), while eukaryotic cells have many organelles.

Photosynthetic Organisms Harness Energy From Sunlight

Photosynthetic organisms can make their own food as well, but they require pigments that capture energy from photons delivered by sunlight. The pigments of photosynthetic prokaryotes are in their cell membrane, while photosynthetic eukaryotes house these pigments in specialized structures called chloroplasts.

Chemosynthetic Organisms Harness Free Energy From Inorganic Compounds

Chemosynthetic organisms, such as bacteria, oxidize inorganic compounds, such as sulfur and methane. This is done in order to make food by combining them with nutrients and converting them into organic matter for energy.

Respiration

The chloroplast and mitochondria are key players in energy conversions (respiration and photosynthesis) in eukaryotes. In autotrophs, the energy products created through photosynthesis are used in cellular respiration. In heterotrophs, the energy needed for cellular respiration is obtained through food. The reactants and products of each cycle are listed below:

1a. Autotrophs only: Photosynthesis in the chloroplast makes glucose and oxygen to be used in cellular respiration:

$$6CO_2 + 6H_2O \rightarrow C_6H_{12}O_6 + 6O_2$$

1b. Heterotrophs only: consume glucose and pass its energy along the food chain

2. Both autotrophs and heterotrophs: Aerobic respiration occurs in the mitochondria, producing water, carbon dioxide, and energy.

$$C_6H_{12}O_6 + 6O_2 \rightarrow 6H_2O + 6CO_2 + 32ATP$$

3. Both autotrophs and heterotrophs: Organisms die, decompose, and their essential elements are re-used, and then the cycle repeats.

Oxidation/Reduction

Before delving into the physiological processes regarding respiration, it is important to have a firm understanding of the concepts of oxidation and reduction, since most biological processes involve a substance that is either being reduced or oxidized. The most important thing to remember is this: oxidation is a loss of electrons, and reduction is a gain of electrons. Below is a helpful mnemonic for remembering this concept:

OIL RIG

Oxidation Is Loss, Reduction Is Gain

For the processes listed below, most electrons are lost by donating a hydrogen atom and the electron that is contains. However, electrons can be lost or gained without hydrogen.

A redox reaction is any reaction in which a chemical is reduced by a complementary oxidation process.

Stage One of Photosynthesis: The Light-Dependent Reaction

Photosynthesis is the complex process that chlorophyll-containing organisms perform to make their own food, which will then be used to create energy through cellular respiration. Chlorophyll is a green pigment responsible for the absorption of a photon (a unit of light), which provides the energy required to begin photosynthesis. Chlorophyll is present in the eukaryotic cells of plants and plant-like protists such as green algae.

Photosynthesis begins with Stage One, also known as the light dependent reaction, which results in the creation of energy in the form of ATP and NADPH. It is light dependent because it requires the energy provided by a photon. It occurs in a complicated series of steps, which are outlined here:

1. Light from the sun (in the form of a photon) strikes a molecule of chlorophyll that is embedded within and around the photosystems lodged within the thylakoid membrane of the chloroplast.

2. The photon excites an electron located in Photosystem II (PSII), which is the first of four protein complexes within the membrane. PSII absorbs photons with a wavelength of 680 nanometers in protein 680 (p680). This excited electron jumps into the primary electron acceptor in the center of PSII, where it is picked up by an electron carrier. The electron serves to transfer energy.

3. Meanwhile, PSII takes a molecule of water and splits it into hydrogen and oxygen, stealing an electron from hydrogen to replace the one it just lost, and releasing oxygen and protons (H^+). This process is called *hydrolysis*.

4. The excited electron travels to the next protein complex, the cytochrome complex — the intermediary between PSII and Photosystem I (PSI) — which uses the energy from the electron

to pump a proton across the thylakoid membrane into the thylakoid space, creating a positive concentration gradient.

5. The electron, having exhausted all of its energy as it moved along the electron transport chain, splitting water and pumping hydrogen ions, travels to PSI and is re-energized by another photon at a wavelength of 700 nanometers in protein 700 (p700).

6. The re-excited electron is picked up by another electron carrier and taken to the NADP+ reductase, an enzyme that uses the energy from the electron to create NADPH, a vessel of stored energy, by accepting hydrogen and two electrons (from the ETC) and donating them to NADP+.

7. Meanwhile, the hydrogen protons that build up within the thylakoid space are propelled by their natural inclination to move away from each other (via the repulsion of their charges) and push their way through the ATP synthase. This uses the energy of the proton gradient to add an inorganic phosphate (Pi) to ADP to create ATP.

8. The ATP generated from the electron transport chain and the electrons carried by NADPH are soon invested in the Calvin cycle to create a high-energy glucose.

The overall net reaction of all of the reactions of oxygenic photosynthesis can be seen in the following formula:

$$2H_2O + 2NADP+ + 3ADP + 3P_i \rightarrow O_2 + 2NADPH + 3ATP$$

The Calvin Cycle

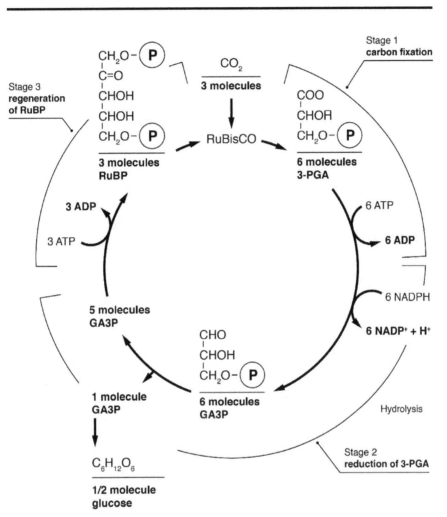

The Calvin cycle is the part of photosynthesis that actually creates glucose. It uses the byproducts ATP and NADPH and the solar energy harnessed in Stage One in a series of events described as follows.

1. Carbon dioxide from the atmosphere enters the plant through stomata on the bottoms of its leaves. It then diffuses into the stroma of the chloroplast, located outside the thylakoid membrane.

2. In the exergonic reaction called carbon fixation, the CO2 then combines with RuBP, a 5-carbon molecule with two phosphate groups, catalyzed by the enzyme called RuBisCO, the most abundant protein in the world. The addition of a sixth carbon causes the molecule to become unstable, so each molecule of RuBP immediately splits into two 3-carbon molecules with a

phosphate group called 3-phosphoglycerate (3-PGA). This process happens three times, resulting in 6 molecules of 3-GPA. (This step is also called the C3 pathway).

1 Turn of the Calvin Cycle: CO_2 + RuBP → 3-PGA

3 Turns of the Calvin Cycle: $3CO_2$ + 3 RuBP → _6 3-PGA_

3. In an endergonic reaction called reduction, NADPH uses energy from ATP to add a hydrogen to each molecule of 3-phosphoglycerate, turning the six molecules of 3-PGA into a 3-carbon sugar called glyceraldehyde 3-phosphate (G3P). ATP supplies energy by donating a phosphate group (P_i) and becoming ADP, while NADPH loses a hydrogen to become $NADP^+$.

    6 ATP + 6 NADPH + 6 3-GPA → _6 G3P_ + 6ADP + $6P_i$ + $6NADP_+$ + $6H_+$

4. Of the six molecules of G3P created, only one is reserved to make sugar, while the other five molecules are reused in an endergonic reaction called regeneration to replace the three used RuBP molecules. It takes two molecules of G3P to make glucose, and since three turns of the Calvin cycle produce only one G3P, this means it takes six turns of the Calvin cycle to make one 6-carbon molecule of glucose. In the following formulas, remember that most of the G3P is used to renew RuBP.

    3 Turns = $3 CO_2$ + 3 RuBP → 6 3-PGA → 6 G3P → 1 G3P exit → ½ $C_6H_{12}O_6$ (glucose)

    6 Turns = $6 CO_2$ + 6 RuBP → 12 3-PGA → 12 G3P → 2 G3P exit → 1 _$C_6H_{12}O_6$ (glucose)_

Photorespiration: C3, C4, and CAM Plants

In certain conditions, plants will halt the production of glucose in a process called photorespiration. C3 plants, the most common, are the most efficient in cool, moist climates. In hot or dry conditions, stomata, the miniscule holes in leaves that enable the transfer of liquid and gases between the plant and its environment, close to conserve water. This can be problematic because it reduces the influx of carbon dioxide. Carbon dioxide becomes scarce, and the oxygen byproduct of the hydrolysis during the light-dependent reaction in PSII builds up, unable to escape. When temperatures increase, RuBisCO has a higher affinity for oxygen and that, combined with the higher O_2 to CO_2 ratio caused by the closed stomata, causes it to bind to O_2 instead of CO_2. This means that carbon cannot be fixed to become glucose, but still it uses ATP to burn up energy, essentially undoing the work of the Calvin cycle. Therefore, photorespiration is resource- and energy-draining, and scientists are not entirely sure of its evolutionary significance. Two alternative systems exist in some plants to avoid photorespiration.

C4 plants and CAM plants are found in tropical and desert climates. They have mechanisms to avoid photorespiration that they will employ if resources are low. They both take an alternative route to the Calvin cycle, so they actually have the same sugar-producing endgame as C_3 plants. These pathways that circumvent photorespiration either require extra energy (as with C_4 plants) or are not as efficient (as in CAM plants) as functioning C3 plants, but the fact that they still produce sugar for life-sustaining energy makes them preferable to photorespiration.

C4 (4-carbon) plants evade photorespiration by using a much more efficient enzyme called PEP carboxylase, which has an affinity for carbon dioxide only. RuBisCO is the preferred enzyme for carbon fixation, since PEP carboxylase requires energy. However, if RuBisCO is being blocked by oxygen in low carbon dioxide situations, PEP carboxylase will bind any circulating carbon dioxide and incorporate it into a C4 product. This 4-carbon product (called oxaloacetate, or OAA for short) releases carbon dioxide to the Calvin cycle to be used by RuBisCO in carbon fixation. ATP is invested to convert it to a 3-carbon sugar that can bind to PEP carboxylase and repeat the cycle. Basically, PEP carboxylase is acting as a carbon dioxide pump to keep levels high, enabling it to make the high energy glucose.

CAM plants, typically found in deserts, conserve water by keeping their stomata closed during the hot part of the day; this prevents dehydration through transpiration. The stomata open at night to capture and store carbon dioxide in organic compounds. Like C4 plants, the carbon is fixed into carbon intermediates. During daylight, these compounds are broken up so that released carbon dioxide can bind to RuBisCO and stimulate sugar production via the Calvin cycle. This pathway is beneficial for plants such as cacti that need to conserve water. It is not as efficient due to uncoupling carbon fixation with the rest of the cycle, but for extremely hot environments, it is adaptive.

Cellular Respiration in Eukaryotes

All organisms, whether autotrophic or heterotrophic, use food to produce ATP in a process called respiration. Cellular respiration is the metabolic process that converts energy from nutrients into ATP and waste products. This will be explained in detail later. Prokaryotes use proteins on their cell membrane to perform respiration, while eukaryotes have specialized structures called mitochondria to do it.

Chloroplasts and mitochondria are the organelles responsible for all energy conversion in eukaryotic cells. All eukaryotes have mitochondria, but only plants and green algae have chloroplasts as well.

Glycolysis

The first step of breaking down glucose to make energy is called glycolysis (literally "glucose-splitter"), and it occurs in the cytosol of cells.

$$C_6H_{12}O_6 + 2ATP \rightarrow 2C_3H_4O_3 + 4ATP + 2NADH$$

Glucose + activation energy $\rightarrow$ 2 Pyruvate + energy + 2 electron carriers

As shown in the previous formula, the overall goal of glycolysis is to break glucose in half and into 2 pyruvate molecules. In doing so, it peels off high-energy electrons that were contained in glucose. Two pairs of electrons (stored in phosphate groups) and two hydrogen atoms are invested into the electron carrier NAD+ that behaves just like the electron carrier in photosynthesis by shuttling electrons from one process to the next.

Glycolysis requires a 2ATP energy investment to proceed to completion, and it produces 4ATP via substrate level phosphorylation. This net gain of 2ATP is a small percentage of the total energy produced in aerobic respiration.

In the absence of oxygen, the 2ATP produced in glycolysis is the only energy gain there is, and fermentation, or anaerobic respiration, will initiate to recycle the electron carrier NAD^+. The two chief types of anaerobic respiration/fermentation are lactic acid fermentation and alcohol fermentation. When muscle cells have exceeded their aerobic capacity, they go into anaerobic respiration, which produces lactic acid and 2 net ATP. Yeast undergoes alcohol fermentation, producing carbon dioxide, ethyl alcohol, and 2 net ATP.

Glycolysis is performed via the following steps:

1. Glucose is converted into Glucose-6-Phosphate (G6P) via the enzyme hexokinase (note that any enzyme ending in -kinase indicates that a phosphate group is going to be donated). Energy is lost because ATP loses a phosphate group and is converted into ADP in the process of creating G6P.

2. G6P is then rearranged, turning from a hexagonal figure to a pentagonal figure, and is configured into Fructose-6-Phosphate (F6P) by an enzyme called phosphoglucose isomerase.

3. F6P is converted into Fructose-1,6-Bisphosphate (F1,6BP) by phosphofructokinase, donating a phosphate group from ATP and turning it into ADP, losing energy.

4. F1,6BP is then broken down into two 3-carbon molecules by the enzyme aldolase. These molecules are DHAP (dihydroxyacetone phosphate) and G3P (glucose-3-phosphate). DHAP acts as a kind of brake on glycolysis, if there is too much energy in the body, this reaction favors DHAP. However, if there is not enough energy in the body, this reaction favors G3P to go on to continue glycolysis.

5. G3P is then converted into 1,3-Bisphosphoglycerate (1,3-BPG) via glyceraldehyde phosphate dehydrogenase. During this conversion, $NADP^+$ is converted into NADPH.

6. 1,3-BPG becomes 3-Phosphoglycerate (3-PG) via the enzyme phosphoglycerate kinase). Energy here is gained in the form of ATP, in which ADP gains a phosphate group through this enzyme.

7. 3-PG is then mutated by phosphoglyceromutase into 2-Phosphoglycerate (2-PG).

8. 2-PG is converted into phosphoenolpyruvate (PEP) via the enzyme enolase. Water is created via this process.

9. Finally, PEP is converted into pyruvate through the enzyme pyruvate kinase, in which energy is again gained by the donation of a phosphate group to ADP to create ATP. Pyruvate is then entered into the Krebs cycle.

The Krebs Cycle

If oxygen is present in a eukaryotic organism, the remainder of the process of respiration will occur inside the mitochondria via the Krebs cycle and oxidative phosphorylation. The main goal of the Krebs cycle is to take pyruvate and break it down, producing NADH and FADH$_2$.

The Krebs Cycle

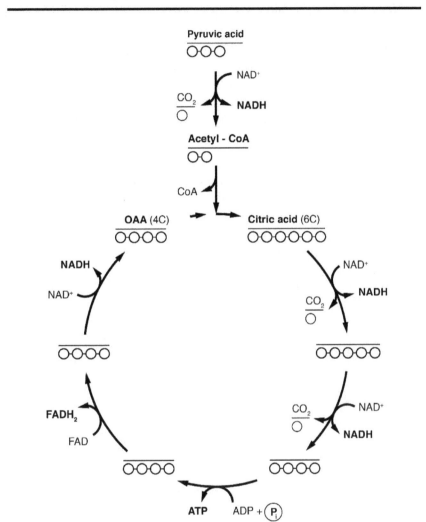

Upon entering the mitochondria, the pyruvate releases a pair of high-energy electrons, and a proton (H$^+$) to the electron carrier NAD$^+$, plus a carbon dioxide molecule. This happens while it is being converted into a two-carbon molecule attached to Coenzyme A, formally called Acetyl CoA. This step is called pyruvate oxidation. Acetyl CoA is used as fuel for the following citric acid (Krebs) cycle.

Including the intermediate stage, where pyruvate enters, two spins of the cycle (one for each pyruvate) produces the following:

- 2 CO$_2$ (from intermediate)
- 2 NADH (from intermediate)
- 4 CO$_2$

- 6 NADH
- 2 FADH$_2$
- 2 ATP (or GTP) via substrate-level phosphorylation. GTP is guanosine tri-phosphate, which is analogous to ATP.

NADH and FADH$_2$ are electron carriers, meaning they donate electrons to the electron transport chain. NADH donates two high-energy electrons and one proton (H+), while FADH$_2$ donates two electrons and two protons (2 H$^+$). These high-energy electron and proton carriers then go through the process of oxidative phosphorylation, where many ATP molecules are made. Using the energy supplied by the electron transport chain, protons are transported through the integral protein complexes I, III, and IV along the inner mitochondrial membrane. The pumping of these protons out of the mitochondrial matrix across the cristae to the inner membrane space, establishes a concentration gradient. Just like in photosynthesis, this gradient provides the proton motive force to generate ATP when the hydrogen ion later passes through ATP synthase, causing it to spin and convert ADP to ATP.

Electron Transport Chain

The energy source for oxidative phosphorylation (also known as the electron transport chain) are the electrons traveling through the membrane of the mitochondrial cristae in a redox reaction-driven electron transport chain. Their high energy is coupled with the active transport of the protons, and as they are passed down the chain in a series of redox reactions, oxygen becomes the final electron pair acceptor. The electrons and the hydrogen ions join an electronegative oxygen to form water.

Energy Extraction from Cellular Respiration

Aerobic respiration uses oxygen and produces 30-32 ATP molecules using the mitochondria. Only a few of the ATP are generated via substrate-level phosphorylation in glycolysis and the Krebs cycle; the vast majority of ATP is generated through the electron transport chain and chemiosmosis.

The exact number of ATP molecules made per glucose molecule varies. Glycolysis and the Krebs cycle each produce a net gain of 2 ATP/GTP via substrate-level phosphorylation. Oxidative phosphorylation is more difficult to calculate. Each electron carrier NADH produces around 2.5 ATP, while FADH$_2$ produces around 1.5 ATP. These are not whole numbers because there is not a direct relationship between electron transport and phosphorylation — they are two different processes. One has to do with electrons traveling down the chain to the final electron acceptor: oxygen. The other has to do with the movement of hydrogen ions. Finally, some of the work done by oxidative phosphorylation might be distributed to other cellular processes because respiration does not exist in a vacuum.

Another example of the flexibility of energy production by respiration is seen in thermoregulation. It was discussed earlier that endothermic organisms have ways to regulate body heat, including shivering and sweating. Another is by using an uncoupling protein in the cristae during hibernation. A mitochondrial protein called uncoupling protein 1 (UCP1) in brown fat cells hijacks the proton motive force by preventing them from entering the ATP synthase and creating ATP. Instead, UCP1 moves the protons by increasing the permeability of the inner mitochondrial membrane, using energy from the proton gradient to be dissipated as heat. This helps keep hibernating animals warm without creating unneeded ATP, which helps animals conserve energy and keep their metabolic rates low.

In addition to two pyruvate molecules produced by glycolysis, six molecules of NADH and two molecules of flavin adenine dinucleotide (FADH$_2$) are produced and used by the ETC. Hydrogen atoms, transported by NADH and FADH$_2$ to the ETC, are used to produce ATP from ADP. The hydrogen atoms form a proton concentration gradient down the ETC that produces energy required to produce ATP. NADH and FADH$_2$ molecules rephosphorylate ADP to ATP via the ETC with each NADH producing three ATP molecules and FADH$_2$ producing two ATP molecules.

Carbon is the Foundation of Biological Molecules

Carbon is the foundation of organic molecules because it has the ability to form four covalent bonds and long polymers. The four organic compounds are lipids, carbohydrates, proteins, and nucleic acids.

- Lipids are critical for cell membrane structure, long-term energy storage, and to help form some steroid hormones such as testosterone and cholesterol.

- Carbohydrates are important as a medium for energy storage and conversion, but also have structural importance. Cellulose (a monomer of glucose) provides structure for plant cell walls. Chitin provides structure for fungi and animals with exoskeletons (such as crabs and lobsters), and peptidoglycan is a carbohydrate/protein hybrid that forms the cell walls of some prokaryotes.

- Proteins are important because enzymes regulate all chemical reactions, but there are also many cell membrane proteins important for structure, transport, and communication.

- Nucleic acids include DNA – the genetic instructions of organisms – and RNA, which is the molecule responsible for turning those instructions into products.

All of these organic compounds require the elements carbon (C), hydrogen (H), and oxygen (O), which enter the food chain through the glucose that it produces through photosynthesis. Some compounds contain phosphorous (P), sulfur (S), and nitrogen (N) as well. Elements such as phosphorous and nitrogen diffuse into the roots of plants from the external environment, are incorporated into organic compounds, and are distributed to other organisms through symbiotic relationships or food webs.

Water's Unique Properties Make it Crucial for Life

While water is not an organic compound (meaning it does not contain carbon), it is a critical molecule for life. While it is covalent, in that the oxygen shares electrons with two hydrogen atoms, it also has polar bonds, meaning that it is slightly charged and borderline ionic. Oxygen has an atomic number of 8, and hydrogen has an atomic number of 1. Think of the oxygen nucleus as having eight positive magnets in it, and the hydrogen nucleus of having one. Now think of hydrogen having one lone electron circling around it. That electron is going to be sucked toward the more electronegative, or "powerful," oxygen nucleus, making the oxygen slightly negative. The hydrogen proton will be exposed on one side, and that part of it will be attracted to any negative oxygen neighbors. This intermolecular attraction between the hydrogen proton and a partially negative neighbor is called a hydrogen bond. Hydrogen bonds are *not* ionic; they are intermolecular attractions. The image of liquid water here shows the hydrogen bonds as dotted lines. These hydrogen bonds give water many unique properties, including the following:

- Cohesion is when water sticks to itself because of the attraction between molecules, as observed in dew.

- Adhesion is when water sticks to other structures, as observed in graduated cylinders when sides of the meniscus stick to the walls.

- High surface tension is when water can support some solids. Solids are supposed to be less dense than liquid, but if there is a large enough body of water, there will be a significant "sticky" film at the top due to the cohesive forces between water molecules. Heavy things can penetrate the film and sink, but lighter objects such as leaves and ice, even though they are solid, will not be able to overcome the attractive forces between the water molecules and will float.

- High boiling point is a result of the great amount of kinetic energy required to overcome water's cohesive forces and vaporize.

- High freezing point is a result of water's attractive forces enable it to be arranged much more easily into the specific hexagonal arrangements that form ice, permitting it to solidify at a higher temperature.

- Solid ice has a very low density due to the space between the lattice/crystalline structure, which explains why ice is less dense than water, as seen in the image below.

Hydrogen bonds

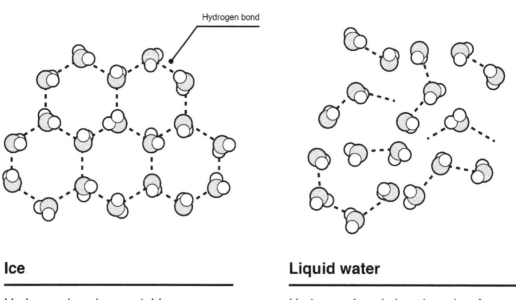

Ice	Liquid water
Hydrogen bonds are stable	Hydrogen bonds break and re-form

These unique properties of water sustain life. Ice floating prevents bodies of water from freezing entirely solid because it insulates the liquid water underneath. This insulation keeps the water below it fluid and allows organisms to survive. Adhesion and cohesion of water in the xylem of plants allows water to travel great distances against the force of gravity.

Cohesion

Cohesion is the interaction of many of the same molecules. In water, cohesion occurs when there is hydrogen bonding between water molecules. Water molecules use this bonding ability to attach to each other and can work against gravity to transport dissolved nutrients to the top of a plant. A network of water-conducting cells can push water from the roots of a plant up to the leaves.

The cohesive behavior of water also causes surface tension. If a glass of water is slightly overfull, water can still stand above the rim. This is because of the unique bonding of water molecules at the surface — they bond to each other and to the molecules below them, making it seem like it is covered with an impenetrable film. A raft spider can actually walk across a small body of water due to this surface tension.

Adhesion

Adhesion is the linking of two different substances. Water molecules can form a weak hydrogen bond with, or adhere to, plant cell walls to help fight gravity.

Water Has High Specific Heat Capacity

Another important property of water is its ability to moderate temperature. Water can moderate the temperature of air by absorbing or releasing stored heat into the air. Water has the distinctive capability of being able to absorb or release large quantities of stored heat while undergoing only a small change in temperature. This is because of the relatively high specific heat of water, where specific heat is the amount of heat it takes for one gram of a material to change its temperature by 1 degree Celsius. The specific heat of water is one calorie per gram per degree Celsius, meaning that for each gram of water, it takes one calorie of heat to raise or lower the temperature of water by 1 degree Celsius.

Water is a Universal Solvent

The polarity of water molecules makes it a versatile solvent. Ionic compounds, such as salt, are made up of positively and negatively charged atoms, called cations and anions, respectively. Cations and anions are easily dissolved in water because of their individual attractions to the slight positive charge of the hydrogen atoms or the slight negative charge of the oxygen atoms in water molecules. Water molecules separate the individually charged atoms and shield them from each other so they don't bond to each other again, creating a homogenous solution of the cations and anions. Nonionic compounds, such as sugar, have polar regions, so are easily dissolved in water. For these compounds, the water molecules form hydrogen bonds with the polar regions (hydroxyl groups) to create a homogenous solution. Any substance that is attracted to water is termed hydrophilic. Substances that repel water are termed hydrophobic.

The Effects of Surface Area-to-Volume Ratios

Organisms require certain reactants to sustain life and have adaptations that maximize access to these reactants. For example, the small intestine contains villi and microvilli, which are millions of small projections that absorb nutrients and deliver them to the circulatory system. The villi and microvilli provide extra surface area to maximize absorption. This is similar to a towel, in that the fluffier the towel the more effective it is at absorption. Root hairs serve the same purpose in plants.

The larger an organism is, the more it needs such structures to maintain adequate levels of material exchange. Single-celled organisms typically remain small to keep a large surface area to volume (SA:V) ratio that will maximize the exchange of materials with the surroundings.

To calculate surface area to volume ratio, simply divide the volume (V) from the surface area (SA):

$$SA/V = Ratio$$

In order to do so, one must consider the shape of the organism for which they are calculating. Most organisms are not just one simple shape, and so calculations for a whole organism will not need to be performed. However, simple shapes, such as cells or bacteria, have a relatively simple calculation.

Here are the common shapes associated with the different kinds of cells, and their corresponding volume and surface area equations:

Shape	Surface Area	Volume
Cube (cuboidal and squamous) 	$6a^2$	a^3
Sphere (spherical) 	$4\pi r^2$ $\pi = {\sim}3.14$	$(4/3)\pi r^3$
Cylinder (columnar) 	$2\pi r^2 + 2\pi rh$	$\pi r^2 h$
Rectangular Prism 	$2\ (wl + hw + hl)$	L x W x H

Cell Membranes Have Selective Permeability

Water is polar because of the bent shape of the molecule — one side of the molecule (the oxygen side) has a negative charge, and the other side (the hydrogen side) has a positive charge. Oxygen, though more negative than hydrogen, has a more positively charged *nucleus,* causing it to pull the two hydrogen molecules' electrons closer to itself. This results in a partial positive charge of the hydrogen side and makes oxygen more electronegative (see the following figure). The molecule is bent because the electrons pulled from hydrogen are not bonded to any other atoms, and are termed "lone" electrons. Further, the two hydrogen atoms on the opposite side of the oxygen are repelled by their own positive forces, and the lone electrons are also repelled by their similar charges. Therefore, they stay as far away from each other as possible while still being held in place by oxygen.

Polar substances dissolve other polar substances. For example, when table salt (NaCl) is placed in water, the molecule is split into Na^+ and Cl^- because Na^+ is attracted to $O2^-$, and Cl^- is attracted to the H^+, essentially "tearing" the molecules apart. Polar substances are said to be hydrophilic (water-loving), and non-polar substance are said to be hydrophobic (water-fearing).

H$_2$O bond

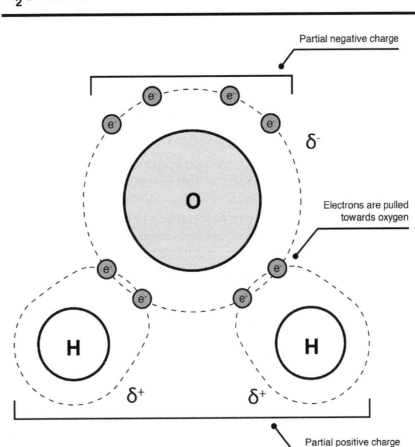

The cell membrane in bacteria and eukaryotes is a phospholipid bilayer that separates the extracellular and intracellular microsystems. It is a dynamic, fluid, and heterogeneous entity commonly described as a fluid mosaic that contains the following structures:

Phospholipid bilayer
The bilayer is composed of polar hydrophilic phosphate "heads" on the exterior and two hydrophobic fatty acid "tails" on the interior. The phospholipid-charged regions are referred to as the heads because

they face the extracellular and intracellular sides. The hydrophobic non-polar tails meet in the middle. There is lateral movement between adjacent phospholipids quite often. Flip-flopping of phospholipids transversely across the membrane from the intracellular to the extracellular side (or vice-versa) is very rare.

Integral and peripheral proteins
Proteins may be completely embedded in the membrane if they are integral, or just attached to one side if they are peripheral proteins. Their placement in the membrane depends on their folding and the polarity or non-polarity of their exposed regions. These proteins can have many different functions.

- Integral proteins can act as channels for movement.

- Enzymes are biological catalysts that speed up chemical reactions.

- Membrane receptors can receive ligands (smaller molecules) and initiate signal transduction cascades.

- Proteins are important for attachment to the cytoskeleton or extracellular matrix.

- Glycoproteins are extracellular carbohydrates attached to proteins and can act as a cell identifier/marker, which is important for cellular recognition. Other markers can be free carbohydrates or carbohydrates attached to lipids (glycolipids).

Cholesterol
Cholesterol is a hydrophobic steroid lipid. It embeds itself in animal cells' lipid bilayers between hydrophobic tails and regulates membrane fluidity. At high temperatures, the embedded cholesterol prevents melting because there is strength in numbers. Squeezing in an extra non-polar molecule provides for many more hydrophobic interactions to contribute to the "glue" that holds it together. In cold temperatures, on the other hand, cholesterol prevents freezing because its ringed structure interrupts adjacent phospholipids. This makes it more difficult for the hydrophobic tails to line up perfectly and freeze, which is similar to how adding electrolytes to water lowers its freezing point.

Plant cell walls are made of a polysaccharide called *cellulose* that offers structural support and protection to the plant. Prokaryotes, fungi, and some protists also have cell walls, although they are not necessarily made of cellulose. Chitin is the structural component of fungi, and bacteria have a cell wall made of peptidoglycan. Cell walls allow movement through channels called plasmodesmata.

All cells contain a cell membrane, which is selectively permeable. Selective permeability means essentially that it is a gatekeeper, allowing certain molecules and ions in and out, and keeping unwanted ones at bay, at least until they are ready for use. This is achieved through active and passive transport, actively allowing molecules and ions through the opening and closing of cell membranes embedded within the phospholipid bilayer (using energy), or passively via a concentration gradient.

The cell membrane, or plasma membrane, has selective permeability with regard to size, charge, and solubility. With regard to molecule size, the cell membrane allows only small molecules to diffuse through it. Oxygen and water molecules are small and typically can pass through the cell membrane. The charge of the ions on the cell's surface also either attracts or repels ions. Ions with like charges are repelled, and ions with opposite charges are attracted to the cell's surface. Molecules that are soluble in phospholipids can usually pass through the cell membrane. Many molecules are not able to diffuse the cell membrane, and, if needed, those molecules must be moved through by active transport and vesicles.

Passive Transport

Passive transport is the movement of substances from high concentration to low concentration without the use of energy. This occurs because of the universe's tendency to achieve a state of equilibrium, or balance. For example, when food coloring is dropped into a cup of water, it spreads to the rest of the water until equilibrium is reached and there is a homologous solution. Passive transport comes in the following three varieties:

- Diffusion: When particles move from high concentration to low concentration, like when hot chocolate powder dissolves in water to form a tasty treat, as seen in the below figure.

- Osmosis: When water moves from high concentration to low concentration via a permeable membrane. In the below figure, there is a high water to solute ratio on the left, and a low water to solute ratio on the right. Water moves to the higher concentration of solutes in order to achieve a more equal water to solute ratio.

- Facilitated diffusion: When particles diffuse through a channel protein due to the membrane's selective permeability

Water is a molecule that travels via facilitated diffusion through proteins called aquaporins. Water movement is determined by its environment, of which there are three types:

- Isotonic environments are achieved when there is a dynamic equilibrium between two solutions. Water will move in and out at equal rates.

- Hypotonic solutions are ones that have a lower solute concentration than the solution they are being compared to. The solution will be low in solute and high in water, so water will move out to achieve equilibrium.

- Hypertonic solutions are ones that have a higher solute concentration than the one they are being compared to. They will be high in solute and low in water, so water will move in to achieve equilibrium.

Animal cells in a hypotonic environment are subject to a net water movement into the cell, so the cell will swell and possibly burst if equilibrium is never achieved. Conversely, hypertonic solutions will cause the cell to shrink as water moves out.

Plant cells are also affected by their water concentrations. The pressure of their cell wall against the cell membrane is called turgor pressure. In hypotonic environments, the cell will swell and the cell wall will press on the cell membrane, resulting in a turgid, or firm, cell. This is healthiest for a plant because it provides the support to hold it upright. Isotonic environments mean that water will diffuse in and out of the plants cells at equal rates. In a hypertonic environment, the plant cell will undergo a process called plasmolysis, where the cell membrane shrinks and separates from the cell wall, resulting in a wilted plant due to lack of turgor pressure.

Protists such as paramecium that live in hypotonic environments will constantly have water moving inward and will never achieve equilibrium. They have a contractile vacuole that actually pumps out extra water using energy.

Water potential (Ψ) always indicates the direction of net water flow. It is determined by the sum of the pressure potential (Ψ_p) and solute potential (Ψ_s).

$$\Psi = \Psi_s + \Psi_p$$

Pressure potential is the force a plant cell wall exerts on its cell membrane (turgor pressure). Solute potential (also called osmotic potential) is defined by the pressure needed to be added to a solution in order to prevent the influx of water into the other solution. It is influenced by concentration, by molarity of the solution, by ionization of the solute, and by temperature of the system. Solute potential decreases as solute is added to a solution, thus the negative sign in the equation below. Distilled water has a solute potential of zero.

$$\Psi_s = -iCRT$$

The equation above shows the relationship between the variables. R is not a variable — it is the pressure constant equal to .0831 L x bar/mole x K. A bar is a very large unit of pressure. The actual variables are as follows:

- First, i is a measure of the solute's ionization. If the solute disassociates, the number of individual particles in the solution increases, which decreases the solute potential. Organic and nonpolar molecules have an i value of 1. Ions have an i value dependent on their disassociation into a cation (a positive ion) and anion (a negative ion). NaCl disassociates into 2 ions (Na^+ and Cl^-) and has an i value of 2. $CaCl_2$ separates into 3 ions ($Ca2^+$ and $2Cl^-$) and has an i value of 3.

- C (concentration) is the molarity (moles per liter) of the solution.

- T is the absolute temperature of the system in Kelvin (°C + 273).

Water passes through the semipermeable cell membrane through aquaporins, while small, uncharged substances, such as oxygen and carbon dioxide, diffuse freely. Water's passage is an example of channel-mediated facilitated diffusion, which is simply diffusing across the cell membrane with the help of a tunnel-like protein. In addition to channel-mediated diffusion, there is also carrier-mediated facilitated diffusion. Carrier proteins, such as the one that allows glucose into the cell, change shape upon solute binding. Facilitated diffusion is a type of passive transport and does not require energy (ATP).

Active Transport

Some transport requires energy for molecules to cross the membrane. Protein pumps are transmembrane proteins that use ATP to change their conformations so that they can force substances against their concentration gradient. For example, the sodium-potassium pump uses ATP to move three sodium (Na+) ions out of the cell and two potassium (K+) ions into the cell. This maintains the steep voltage gradient across the membrane in neurons and muscle cells so that the intracellular environment is more negative than the extracellular one. This membrane potential, along with the concentration gradient, creates an electrochemical gradient.

Another type of active transport, the proton pump, is critical to photosynthesis and respiration through its maintenance of proton gradients across the thylakoid and cristae, respectively.

Endocytosis and exocytosis are examples of active transport required for the bulk movement of substances. They require vesicles, which are plasma membrane-bound delivery sacs. The different types of endocytosis are all similar in that the membrane folds inward so that it pinches off to create a vesicle that carries substances into the cell. The contents then travel to the lysosome, which is an acidic organelle that contains digestive enzymes. The lysosome breaks down the materials so that the cell can use them. Types of endocytosis include:

- Phagocytosis, or "cellular eating," occurs when a cell engulfs large particles and internalizes them by using vacuoles. This only happens in specialized cells such as immune cells.

- Pinocytosis, or "cellular drinking," occurs when a cell engulfs droplets of extracellular fluid and surrounds them with vesicles. This happens routinely in animal cells.

- Receptor-mediated endocytosis occurs when a ligand binds to a receptor protein that initiates a signal transduction cascade.

Endocytosis

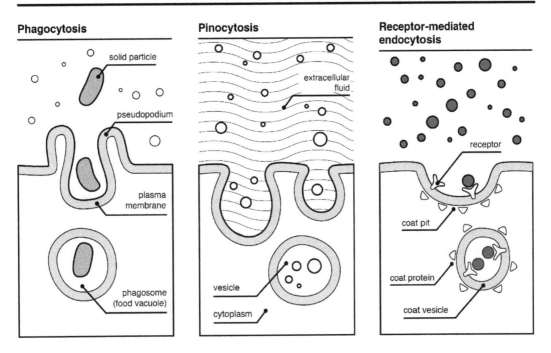

Exocytosis is the excretion of substances via vesicles. Neurotransmitters are excreted via exocytosis at synapses between neurons. Vesicles also release digestive enzymes, hormones, and cellular wastes.

Exocytosis is also used when membrane proteins processed in the Golgi apparatus attach to vesicles that deliver the proteins to their membrane destination.

Exocytosis

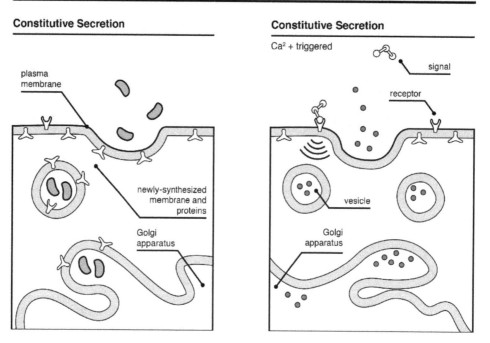

Eukaryotic Organisms Contain Organelles

Prokaryotes and eukaryotes have many structural differences, due to the fact that eukaryotes have evolved in a way that involves a "divide and conquer" mentality. Eukaryotes have specialized structures that prokaryotes lack, and this division of labor makes many metabolic reactions more manageable and efficient. The specialized structures unique to eukaryotes are called "organelles."

Eukaryotic organelles have the following characteristics:

- Chloroplasts are the sites of photosynthesis in autotrophs.

- Mitochondria are the sites of aerobic respiration.

- The nucleus is the DNA-containing structure surrounded by the nuclear membrane.

- The nucleolus is responsible for RNA synthesis.

- The ER (endoplasmic reticulum) is composed of stacks of membranes known as cisternae.

- The smooth ER does not contain ribosomes and is responsible for lipid synthesis.

- The rough ER contains ribosomes that synthesize membrane-bound proteins. As the protein travels from the cis face (nucleus-facing) to the trans face (Golgi-facing), protein folding occurs with the help of chaperonin proteins. Upon completion, the ER delivers this protein to the Golgi.

- The Golgi body is also referred to as the Golgi apparatus or Golgi complex. It is similar to the ER in that it is also series of membrane sacs, but is different in that its job is to modify, sort, and deliver proteins via vesicles to their destination.

- Vesicles are membrane-bound sacs that deliver membrane proteins to their destination and molecules to lysosomes.

- Peroxisomes are detoxification stations that convert the harmful byproduct hydrogen peroxide into water and oxygen using the enzyme catalase. They also break down long chains of fatty acids to be used in other cellular processes.

Differences Between Animal and Plant Cells

Animal cell

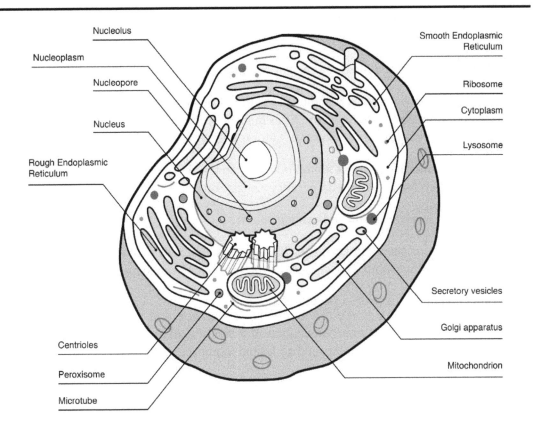

Animal and plants cells differ in a number of ways. Like the chloroplast, there are also other structures that only some cells contain, including the following:

- Animal cells have enzyme-rich lysosomes. After endocytosis, vesicles deliver contents to the lysosome so that they can be broken down and recycled.

- Animal cells have centrosomes that contain two centrioles, which play an important role in mitosis. They contain microtubule centers that elongate animal cells and separate chromatids during cell division.

- Plants have large central vacuoles that store water and food. Some protists have contractile vacuoles to maintain the right osmolarity, a measure of solute concentration. Animals have food vacuoles that contain the contents of phagocytosis.

- Cilia are specialized hair-like projections on some eukaryotic cells that aid in movement, while flagella are long, whip-like projections that are used in a similar capacity.

- Some animal cells have projections called villi and microvilli (such as those in the intestines) that serve to increase surface area and provide the cell with more opportunities to interact with its environment.

Prokaryotic Structures

While prokaryotes are a fraction of the size and missing many of the eukaryotic structures, they do have a lot in common with eukaryotes.

- The plasma membrane surrounds cells. All cells, with the exception of animal cells, have a cell wall.

- The cytosol is the fluid that holds all cellular components.

- The cytoskeleton in eukaryotes is composed of microtubules, intermediate filaments, and microfilaments and serves to hold structures in place. The eukaryotic cytoskeleton also transports and direct vesicles. Prokaryotic cytoskeletons are much simpler but do exist.

- Bacterial cell membrane proteins are homologous to the thylakoid and cristae, proteins in eukaryotes, which allow them to do photosynthesis and respiration.

- Ribosomes perform translation and make proteins.

- Motile bacteria, unicellular protists, and specialized animal cells such as sperm have a complicated flagellum, which is a whip-like structure that aids in movement.

Bacteria even have the following unique structures.

- Pili in bacteria allow for bacterial conjugation (transfer of genetic material between cells, a form of reproduction), or when bacteria send a plasmid to another cell.

- Fimbriae are sticky projections from bacteria that help them stick to hosts or each other.

- Plasmids are circular pieces of DNA that carry at least one gene. Pili actually extend from one bacteria to another, connect to the neighboring bacteria, and inject a plasmid. This process of conjugation contributes to the bacteria kingdom's diversity.

Feedback Mechanisms Help Regulate Fluid Balance

Organisms regulate cellular processes through feedback loops to fine-tune many other processes, including water osmolarity.

In blood, osmolarity refers to the concentration of the collective solutes and water in the blood, and it is regulated by the hypothalamus. The hypothalamus is the bridge between the nervous and endocrine systems via the pituitary gland, and it contains osmoreceptors that sense blood osmolarity. If blood has a low percentage of water, then an individual is dehydrated and has high blood osmolarity. If osmolarity is high, the hypothalamus stimulates the pituitary gland to release stored antidiuretic hormone (ADH). ADH stimulates the kidneys to increase water permeability in the collecting ducts, which increases reabsorption of water and reduces urine volume, meaning less water is lost through urination. This water-retaining mechanism will eventually lower the osmolarity, and when it falls below its set point, osmoreceptors will inhibit the hypothalamus from initiating reabsorption, as illustrated in this diagram:

This diagram is an example of a negative feedback loop because the stimulus feeds back to the regulator in order to change the production in the opposite direction. If production is too high, the system turns off, and if production is too low, the system turns on. Negative feedback loops respond to environmental conditions to keep them at a "set point."

Positive Feedback Loops Also Help Maintain Homeostasis

Positive feedback loops also exist to amplify responses in the presence of production, such as with intracellular signaling cascades or recruitment of cells in the immune system. They do not necessarily ensure cellular homeostasis, but rather they ensure functionality. For example, the production of oxytocin during childbirth is a critical positive feedback loop. In the case of childbirth, the initial stimulus is pressure on the cervix, which sends a message to the brain that causes the pituitary gland to secrete oxytocin, which acts on the uterus to stimulate contractions. As the pressure on the cervix increases, more oxytocin is produced in a positive feedback cycle. Once the baby and placenta are delivered, pressure on the cervix disappears, the pituitary stops producing oxytocin, and uterine contractions stop. Without this positive feedback loop, it would be impossible to provide the force necessary to deliver babies vaginally.

Malfunctions in Feedback Loops Can be Fatal

Non-functional feedback loops can be deadly. The classic example is type I diabetes, in which there is an autoimmune response against the beta cells of the pancreas, the insulin-producing cells.

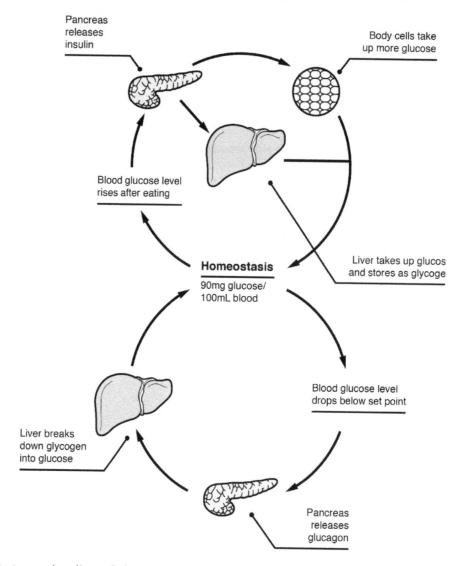

The balance between insulin and glucagon is disrupted when the pancreas can't produce insulin. Insulin serves to open the carrier proteins on cellular membranes, allowing glucose entry. In the absence of insulin, the carrier proteins in cells are not signaled to take in glucose, causing blood sugar to rise. In normal, healthy conditions, in response to high blood sugar, the pancreas is stimulated to produce more insulin from the beta cells (specialized insulin-producing cells). In diabetics, the beta cells have been destroyed by the autoimmune system. The whole bottom portion of the feedback loop is bypassed, and as a result, diabetics become severely hyperglycemic without artificial insulin administration. Severe hyperglycemia can result in coma and death.

Organisms Respond to Environmental Changes

In addition to obtaining energy and maintaining internal homeostasis, organisms must be able to respond to stimuli in their environments in order to ensure survival. The classical experiment to demonstrate phototropism — the physical orientation of a plant in response to light — is a key example.

Photoperiodism and Phototropism in Plants

Rapidly producing cells in the tip of the stem normally produce the hormone auxin, which is responsible for stem elongation. The cells that are exposed to sunlight suppress auxin, causing the shadowed-side to grow at a relatively higher rate. This causes plants to bend toward light, ensuring maximal solar exposure for a plant's leaves.

Another example of the environment affecting an organism is the "fight-or-flight" response in humans. This response is activated in times of stress and is named for the two primitive responses to danger:

- Fight
- Run away as fast as you can

Either option requires energy, strength, and speed, and a pair of hormones secreted by the adrenal glands help achieve just that.

Adrenal glands near the kidney secrete a pair of hormones called norepinephrine and epinephrine, also known as noradrenaline and adrenaline, in response to a stress perceived by the brain. If a human encounters a stressor such as a gorilla or an impending AP Biology exam, the adrenal glands secrete these hormones. These hormones divert blood flow from the skin, kidneys, and organs involved in digestion. The blood is instead rerouted to cells in parts of the body involved in the fight-or-flight response, since they need maximum cellular resources.

Increased blood flow to the muscles, liver, and fat cells stimulate the breakdown of carbohydrates and fats, providing instant energy. Blood circulation is also increased in the heart and lungs, increasing pulse and dilating bronchioles, two actions that increase the rate of cellular respiration. The brain also has increased blood flow, which stimulates quick problem-solving and coordination.

Abiotic Factors Affect Cell Activities

Abiotic (meaning "non-living") factors that affect organisms include sunlight, terrain, precipitation, and temperature, among other things. Many organisms have evolved so that they have adaptations to deal with abiotic limitations, such as the Venus flytrap and pitcher plant. These carnivorous plants are photosynthetic and are not actually eating their captives. Instead, they are capturing bugs to attain their nitrogen. These plants evolved in areas that had nitrogen-poor soil, and their animal-like tendencies are due to abiotic pressures.

An example of an abiotic factor's effects on cellular processes is the relationship between the changes in seasons and the pigment changes in deciduous trees. The shorter days and temperature changes stimulate cork cells to grow between the trees' leaves and stems, blocking water delivery to the leaves. They do this to prevent water loss through transpiration that occurs in the stomata, which cannot be regained because external water sources freeze in the winter. The shedding of leaves reduces the total surface area through which evaporation can occur. The water blockage to the leaves prevents chlorophyll regeneration, and the other pigments become visible. The yellow, orange, and red xanthophylls, carotenoids, and anthocyanins are then able to display their beautiful fall colors.

Biotic Factors Also Affect Organisms

Biotic interactions are important as well. Normal cells have the ability to stop dividing if they have reached capacity, but this density-dependent interaction is lost in cancer cells, leading to uncontrolled proliferation. Even bacteria respond to their environment by quorum sensing, in which they react to bacterial density by signaling neighboring cells. If an optimal density is reached, they produce a slimy biofilm that aids the population in gaining nutrients. On the organism level, there are many communal relationships, including competition, predator-prey, and symbiosis. Symbiosis is cohabitation of two organisms of different species, and there are three types:

- Mutualism: Both organisms benefit from the relationship, such as ants and aphids. Aphids provide a sugary sweet substance that ants eat, while ants provide aphids protection from predators. Bees and flowers are another example. Bees get nectar, while plants are aided in pollination.

- Commensalism: One organism benefits and the other one has a neutral reaction – neither helped nor harmed. For example, barnacles attached to whales are able to ride with the whale into plankton-dense waters, making food more accessible. The whales also provide protection from predators, since the barnacles are more likely to be eaten while stationary. Whales don't benefit from the barnacle attachment, but nor are they hurt by it.

- Parasitism: One organism benefits while the other is harmed. The parasite drains nourishment from its host, slowly and continuously. Some examples are tapeworms, fleas, and ticks.

Organisms are also affected by the availability of food. Food webs can be completely transformed if a particular population changes drastically.

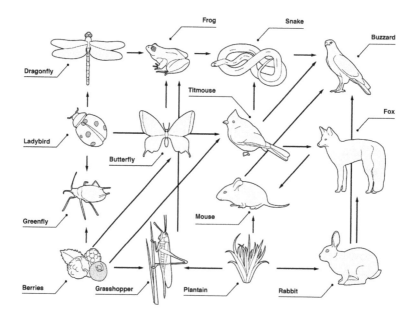

If trees or other plant food sources are destroyed due to deforestation, the primary consumer population will decrease dramatically, as will every trophic level above it.

In marine ecosystems, algal blooms have an opposite effect. The *proliferation* (rapid reproduction) at the producer level takes over and destroys the ecosystem. The algae consume the resources of the ecosystem, producing toxic wastes, causing other species to die and rot. The environment then becomes rich with bacterial decomposers. The bacteria accumulate and use all the available oxygen, and other organisms suffocate.

Homologous Homeostatic Mechanisms are the Result of Common Ancestry

The standing theory is that all organisms come from a common ancestor, due to the staggering amount of homology between all living things. For example, organisms as diverse as archaea, bananas, amoebas, and humans all have the exact same structure and function of DNA, as well as conservation in the replication and transcription of it. Eukaryotes have three RNA polymerases and bacteria have only one, but they work in similar ways. This homology and conservation across all organisms suggests a single common ancestor (called the LUCA, or last universal common ancestor). The phylogenetic tree, also called the evolutionary tree, is a branching diagram that illustrates the relationships between species and their ancestors in a tree-like form, right down to the LUCA.

Plants are a kingdom of organisms that have several common characteristics and therefore, a relatively recent common ancestor. They are all photosynthetic and they all have chloroplasts; however, their methods of obtaining carbon dioxide and water are quite different between the terrestrial and aquatic varieties. This demonstrates both their homology and their divergence along the phylogenetic tree due to adaptations and natural selection.

Gas exchange and water delivery in terrestrial plants are regulated by the stomata, pores with surrounding guard cells that regulate opening and closing. These stomata underneath the leaf are important for many reasons. One is that when water evaporates from the stomata, the movement acts like a siphon, providing the force to pull up water from the roots like a straw. Another is that stomata conserve water by "closing the straw" when water is scarce, so it isn't lost. If water is plentiful, the guard cells fill with water and expand, opening the pore and facilitating evaporation from the stomata in a process called transpiration. In the absence of water, the guard cells close and prevent transpiration. These pores are also the sites of gas exchange in terrestrial plants.

Plants submerged in water do not contain stomata and have no evolutionary advantage for adaptations regarding water conservation. They obtain carbon for photosynthesis via the CO_2 gas that is dissolved in water. Even though terrestrial and aquatic plants have much in common, this single structure – the stomata – illustrates a split in the phylogenetic tree.

Homology is Demonstrated in Nitrogenous Waste Production and Elimination

The animal kingdom also has diversity peppered by varying degrees of homology. This can be observed in many ways, such as by these examples and the variety of mechanisms for osmoregulation and the removal of nitrogenous waste in animals.

Animal	Excretory Structure	Excretory Mechanism
Flatworm	Protonephridia	Inner closed tubules surrounded by cilia-containing flame bulb caps draw in external fluids that travel through the tubule. Fluids are processed, then excreted at the other end of the tubule.
Most annelids and earthworms	Pair of Metanephridia	Collecting tubules (open-ended on either side) contain a ciliated opening to collect fluid and deliver it to the storage bladder, which open and excrete waste to the outside. Capillaries surround the collecting tube.
Insects	Malpighian tubules	Long closed tubes that project into an open circulatory system. Fluid travels to the rectum as water is reabsorbed along the way. Nearly dry waste is excreted with feces from the rectum.
Vertebrates	Kidney composed of nephrons	Nephrons have ascending limbs and descending limbs surrounded by capillaries. The tubules end with a collecting duct that empties to the bladder. The nephron also has a proximal tubule, descending Loop of Henle, ascending Loop of Henle, and a distal tubule preceding the collecting duct.

These diverse structures have a large amount of homology as well, aside from the fact that they have the same function. They all have tubes. Some have a waste storage unit. Some have cilia containing openings. Due to evolution, there is an undeniable conservation across the animal kingdom.

Physiologic Responses to Toxins

All plants and animals have some sort of innate immunity. Even bacteria have restriction enzymes that have evolved as a primitive defense against viruses. These are enzymes that recognize specific palindromic sequences of DNA along a molecule and cleave them at specific sites, dissecting them into smaller, non-harmful fragments.

Lysozymes, enzymes with natural antibiotic properties, protect against foreign invaders by damaging bacterial cell walls. They are found in many secretions, including tears, saliva, mucus, milk, and even egg whites. Insects have lysozymes in their digestive tracts that protect them from disease. Their exoskeletons made of chitin also serve as an effective barrier against foreign invaders. Should a pathogen evade these defenses, hemocytes travel in the hemolymph, the insect circulatory system, and ingest alien particles via phagocytosis. Some hemocytes trigger production of chemicals that kill parasites, bacteria, and fungi.

Mammals have an advanced immune system. Skin epithelial cells, as well as the epithelial lining of the inner mucous membranes, provide protective physical barriers. Mucus traps microbes and washes them away, along with saliva and tears. These physical defenses are enhanced by innate chemical defenses as well. Mucous, tears, and saliva contain lysozymes. The stomach, oil glands, and sweat glands produce acidic fluid that is hostile to many pathogens. Another non-specific chemical defense is stimulated by infected cells themselves. They release chemicals called interferons that provide a localized alarm to surrounding cells. These signals stimulate neighbors to produce substances that inhibit viral replication. Some interferons recruit and activate white blood cells too.

Inflammation

The inflammatory response is also nonspecific and part of the innate defense of mammals. It is characterized by heat, swelling, histamine release, redness, pain, and white blood cell recruitment.

At the site of an injury — a cut, for example — pathogens are released and cells called mast cells release histamine, which causes the dilation of capillaries, allowing them to become more permeable and to increase blood flow to the area. This results in redness and swelling. Clotting elements then move from the blood to the injury, while a tiny army of white blood cells are recruited to the site due to increased circulation and are attracted to the injury via chemical signals called cytokines released by macrophages. The white blood cells, primarily neutrophils, phagocytose ("eat") the pathogens, allowing the wound to heal. It is very common to feel pain during the inflammatory response, as the pressure caused by swelling stimulates nerves.

Other white blood cells play a role in this non-specific immune response.

- Dendritic cells are phagocytic cells that reside in tissues.

- Eosinophils are in mucous membranes and detect multicellular parasites, such as worms, and secrete toxic enzymes to destroy them.

- Natural killer cells circulate and identify viral-infected and cancer cells and secrete toxins to destroy them.

This highly advanced homeostatic system has another component. If a pathogen is still not destroyed, there is a complicated specific immune response that will be discussed in Section 2.E

Pain, though deeply uncomfortable, is beneficial. It alerts the organism that there is an issue and helps to prevent further damage. For example, an animal that has a broken leg will not be able to run on it, which would damage an already vulnerable bone. A cat that licks its wounds does so in order to stimulate blood clotting. Without pain receptors, these adaptive responses wouldn't happen.

Fever, an increase in body temperature as a result of the immune system's defense against infection, helps to destroy bacteria and viruses that are sensitive to temperature change, as well as increase the number of lymphocytes capable of killing pathogens. The benefit of a fever may also be to increase enzymatic reactions to stave off infection. This is demonstrated in the following graph. As temperature increases, the kinetic energy of substrates and enzymes increase, collisions happen faster, and reaction rates increase. If temperature exceeds a certain point, however, it can be very dangerous because enzymes begin to denature and completely lose functionality.

Ecosystem Disruptions Impact Homeostasis

Ecosystems also respond to environmental threats, but unfortunately, they do not have mechanisms to defend themselves as organisms do. Should a natural disaster such as a hurricane, flood, volcano, or fire destroy an ecosystem, its only recourse is to recuperate by rebuilding with the surviving seeds and spores. Secondary succession will occur as another dynamic ecosystem emerges, one that may be similar to the one it replaced, but it will not be identical.

Immune Responses in Plants

The last section discussed animal innate immunity, but plants have defenses as well. Physical defenses, such as thorns and poison, are defenses against consumers.

They also have chemical defenses against microbes. It is not likely that plants will ever develop immune systems as complex as those in animals, since it requires a costly energy investment. However, plants that have no immune system are unlikely to survive. Therefore, plants have co-evolved with avirulent strains of pathogens, which are much less harmful than virulent strains. Harmful viruses would decimate plants due to their weak immune systems, and in doing so they would quickly destroy their host and then be homeless and starving. Less-damaging plant pathogens are, therefore, the norm, since they enable both the plant and the parasitic organism to survive.

Avirulent pathogens have protein effectors (Avr genes) that cause infection in plants that lack the specific resistance (R) protein, a gene responsible for resistance against pathogens. If the R protein is present and the pathogen effector protein binds to it, it initiates a signal transduction cascade that mounts a strong immune response. Part of that response is called the hypersensitive (HR) response, a localized general chemical defense that kills cells surrounding the site of infection. Additionally, modification to the surrounding cells' walls prohibits spreading of the pathogen. There is also a distal immune response; the dying cells secrete methylsalicylic acid that is delivered to non-infected areas and converted to salicylic acid, which signals a systemic, or "whole-plant," immune response.

This description of a plant's general response is akin to innate immunity in animals. Mammals also have an adaptive, or specific, immune response. The cells involved in the adaptive immune response originate from bone marrow and are called B lymphocytes and T lymphocytes. The ones that mature in the bone marrow are B cells and the ones that travel to the thymus to mature are the T cells.

Antigens are anything that activates B and T cells by binding a small region called an epitope to one of their antigen-specific receptors. This may include pathogens, such as bacteria or viruses, toxins or foreign cells introduced by transplantation, or even an organism's own cells in autoimmune responses.

B cells and T cells are different in structure and function, so antigen binding is different between them.

Upon an epitope binding to a B cell's binding site, the B cell proliferates and its daughter cells secrete its characteristic antigen receptor. This antigen receptor is referred to as an antibody or an immunoglobulin (Ig). T cells behave differently in that they only recognize host cells that present fragments of a pathogen's antigen after ingestion.

There are millions of different B and T cell receptors, and this diversity is due to the many possible combinations of a transcribed immunoglobulin gene. The gene contains light and heavy chains of the B cell receptors, and each one can be arranged in several different structures due to recombinase, an important enzyme in antibody development.

These lymphocytes are involved with two different immune responses: cell-mediated and humoral. The cell-mediated response involves T cell destruction of host cells, as outlined below.

- An antigen-presenting dendritic cell, macrophage, or B-lymphocytes engulf the pathogen and digest it. The pieces of digested antigens are displayed on MCH (major histocompatibility complex) class II on the surface of antigen-presenting cells. MCH class I complex contains the host cell's peptides, allowing lymphocytes to recognize them. This ensures that the immune system does not attack its own cells when the antigen is not being presented, and is seen on all host cells. The white blood cells with both kinds of MHC molecules are the specific activators of the helper T cells.

- Helper T cells bind to the antigen and stimulate the humoral and cell-mediated response via the cytokine release.

- Cytotoxic T cells are activated by helper T cell signals.

- Cytotoxic T cells recognize MHC class I molecules on host cells and secrete proteins that trigger cell death.

The humoral immune response includes B cell activation and involves antibody neutralization of pathogens in the circulatory and lymphatic vessels. It begins the same way as the cell-mediated immune response with antigen presentation and helper T cell signaling. In this response, B cells proliferate into memory B cells and effector cells called plasma cells that secrete antibodies.

Antibodies are tags that mark invaders for destruction. They neutralize surface proteins of a pathogen, preventing them from binding to and affecting its host cell. Antibodies can also cause apoptosis (programmed cell death) in infected body cells. Host cells that present epitopes recruit antibodies, which recruit natural killer cells.

Cell Death

Cells die via two different mechanisms: necrosis and apoptosis. Necrosis is involuntary cell death, in which the cell is damaged via external forces, such as an injury, exposure to toxins, or lack of oxygen, and usually is the cause of the inflammatory response. Apoptosis, however, is when the cell essentially kills *itself* when it is no longer needed, by breaking itself down and then being engulfed by macrophages. Apoptosis is most common in embryonic and fetal development — for example, the webbing in-between the toes and fingers of the fetus in the womb, which gets broken down via apoptosis before the baby is born. These cells respond to signals in the body that instruct them to commit suicide.

B cell activation not only produces antibody-making cells, but it also produces memory B cells that keep circulating and are not transient. They remain behind and, should the pathogen be encountered again, they immediately recognize the invader and divide quickly to produce many more effector cells. This results in a much faster and stronger secondary immune response because there is no lag time while B cells are proliferating into plasma cells. Vaccines manipulate the immune systems of animals by delivering inactive pathogens that enable these memory cells to develop.

Gene Expression Affects Observable Cell Differences

Mitosis is the process of cell division and replication in which the cell (mother) produces two identical copies of itself into daughter cells (except in rare mutations). This occurs because the DNA within the cell's nucleus is cleaved into two equal sets of chromosomes — the gene-carrying strands made of nucleic acids and proteins. Mitosis occurs throughout an organism's lifetime, producing cells during growth and development, and also replacing those lost through injury or apoptosis.

If all cells in the human body are identical, how can there be such specialization? For example, how can a heart and a stomach have identical information if they appear so different? The answer lies in the expression of different genes to form the countless different types of cells that make up an organism.

Organisms contain a *genotype* and a *phenotype*. The genotype is the DNA present in the cells that code for the genes, and the phenotype is the set of observable traits that are expressed. For example, a brown-eyed girl may have genes for both blue and brown eyes, but the actual physical trait expressed is brown eyes. So the genotype is blue and brown eyes, but the phenotype is brown eyes.

Human DNA are long strings of protein-based pairs with genes peppered along the 46 chromosomes. Chromosomes are comprised of tightly coiled proteins, RNA, and DNA and are collectively called chromatin in eukaryotic cells. As cells differentiate, chromatin is divided into euchromatin and heterochromatin. Euchromatin is packed less tightly than mitotic chromosomes, while heterochromatin remains densely packed. Euchromatin has open regions of chromosomes, while portions of the DNA become silenced and closed in heterochromatin. Different tissues have different portions of the genome open for transcription, which explains why different cells have different phenotypes. Differential gene expression creates tissue-specific protein combinations that tailors cell structure and streamlines cell function, making the whole organism more efficient.

Just as fetal development occurs in the mother's womb, it is the mother's egg that is responsible for initial development of the zygote (the fertilized ovum). There are specific molecular changes that occur in early divisions that induce cells to differentiate in predictable ways. The location of cytoplasmic determinants, including messenger RNA (mRNA) and organelles, are the cues that set things in motion. Cell-to-cell localized communication also plays a role in early developmental differentiation.

One effect of determination is pattern formation, or the spatial organization of an organism. Dorsal/ventral (back/front) and anterior/posterior (head/tail) structures don't align by chance. Maternal effect genes, also known as egg-polarity genes, establish an axis that essentially directs the developing cells on which is the front end and which is the back end. If mutant offspring lack them, they do not develop properly and may be lethal to the embryo.

Consider the gene *bicoid*. A drosophila (fruit fly) embryo lacking a functional bicoid protein will have two posterior ends and no head. Normally, bicoid mRNA is concentrated at the anterior end of the unfertilized egg; post-fertilization, the diffusing bicoid creates a gradient with the protein concentrated at the anterior end, dilute at the posterior end. This gradient is important for the development of anterior (head) and posterior (anus) ends. Bicoid is just one of many homeotic genes that are responsible for early pattern formation. Drosophila geneticists have even been able to generate flies with legs in place of antennae due to a mutation of a single developmental gene.

Mammalian transplant experiments have shown that organism development can be surprisingly conserved. In 1997, Dolly the sheep was born. She was the first cloned mammal. The diagram below illustrates how the DNA from a donor sheep, switched out with the nucleus of a different sheep's egg, developed into a clone. No fertilization was required because the donor nucleus had the full set of chromosomes. Upon transplantation of the nucleus, the cells went through development as if they were naturally developing when placed inside a carrier mother.

This can be seen here:

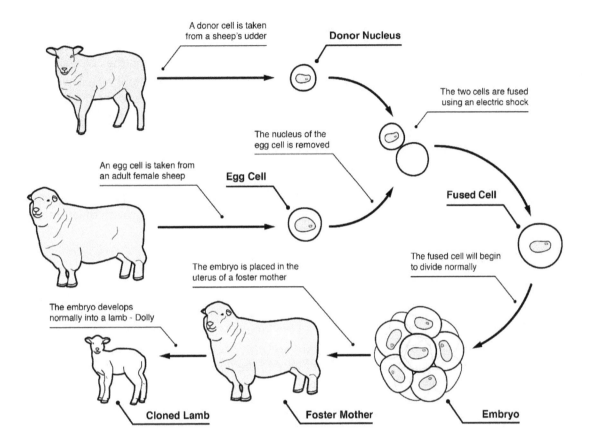

MicroRNAs (miRNAs) and small interfering RNAs (siRNAs) are small RNA segments capable of controlling the expression of genes either by degrading transcripts or blocking translation. These also may play a significant role in organism development and differentiation. RNA interference is a tool used by scientist that exploits these miRNAs by transfecting them into cells, blocking expression of a particular gene of interest, then observing functional effects. Nematodes (roundworms) have been an important model because scientists have been able to silence expression of about 86% of their genes, one at a time, to study gene function.

Plants have a specific series of events in their development as well. Pre-germination seeds are surrounded by food and a thick seed coat. The seed can be dormant for years, but in the plant's optimal temperature and in the presence of water, germination will occur. Post-germination development is affected by gravitropism and phototropism. Gravitropism ensures that roots will always shoot down and stems will always grow up, regardless of position. Phototropism results in plants growing toward sunlight.

Phototropism is a Response to the Environment

Phototropism is a plant's response toward light. It works via the hormone auxin, which causes stem-tip cells to elongate in the absence of light due to differential auxin expression. Auxin travels down the stems, from the tip to the base, in a unidirectional manner dubbed polar transport. This is not a gravitational phenomenon; an upside-down stem has the same effect.

Auxin stimulates cellular growth only if expressed at a specific concentration. If concentrations exceed a certain threshold, it may even inhibit elongation. The acid growth hypothesis postulates that auxin stimulates proton pumps in the cell membrane, causing an acidic environment in the cell wall that activates enzymes called expansins. These proteins weaken the cellulose in the cell wall. The cell wall then expands due to its weakened state and the increase in turgor pressure due to the increased uptake of water by altering membrane polarity with proton gradient. Auxin also has an effect on the transcriptional level and causes the increased expression of genes that aid in this process.

Auxin also affects branching patterns. If a branch reduces its production of auxin in a feedback loop of sorts, it indicates that the branch is a low-level producer. It will then stimulate growth of lateral buds below the branch.

Photoperiodism is Response to Daylight

Photoperiodism is a plant's response to the relative lengths of day and night. Unlike phototropism, it is the amount of darkness that determines blooming. For example, flowers that bloom in the spring are not responding to the lengthening days, but instead are influenced by the shorter nights. "Short-night" plants will not bloom if a continuous amount of darkness exceeds a threshold, as shown in the following top two images. The top one in the figure does not exceed the dark threshold, so it blooms. The second scenario exceeds the dark threshold, so it doesn't bloom. The third scenario shows a situation where the darkness has been interrupted by a flash of light, so the flower blooms because it did not surpass the dark threshold.

Short-day plants, like the poinsettia, flower outside of the spring and are dark-sensitive as well. If periods of darkness are interrupted by light, then flowering is inhibited because they do not have any dark periods that surpass the night threshold. Some plants are so sensitive that even one minute of darkness-interrupting light exposure will inhibit flowering.

Some plants, on the other hand, are day-neutral and are completely independent of the light/dark cycle. Many plants are also responsive to temperature and will not flower unless exposed to a specific temperature for a certain amount of time.

Here's an illustration of that:

Photoperiodic control of flowering

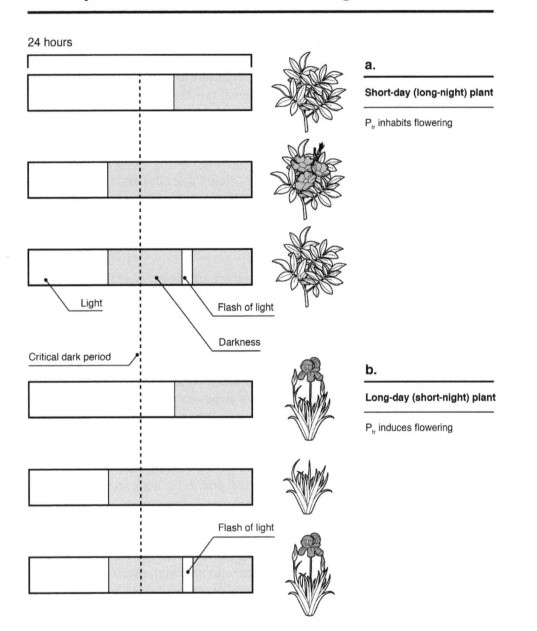

24 hours

a.

Short-day (long-night) plant

P_{fr} inhabits flowering

Light

Flash of light

Darkness

Critical dark period

b.

Long-day (short-night) plant

P_{fr} induces flowering

Flash of light

Animals Respond to Environmental Cues

Animals also respond to environmental cues in a synchronous manner. Circadian rhythms are the predictable animal responses to 24-hour days and are partially controlled by melatonin, a hormone that aids in sleep. High melatonin expression occurs in winter, a period of long nights.

Women's menstrual cycles are also cyclical and regulated by tight hormone control. The cycle begins with the release of a hormone called gonadotropin-releasing hormone (GnRH) by the hypothalamus, which causes the anterior pituitary to secrete low levels of FSH (follicle-stimulating hormone) and LH (luteinizing hormone). FSH stimulates follicular growth in the follicular phase, and the follicles begin to produce estradiol. Oocytes, or eggs, mature, and estradiol feeds back to the hypothalamus/pituitary, inhibiting FSH and LH expression. Pre-ovulation, estradiol spikes to stimulate the hypothalamus/pituitary so that GnRH, FSH, and LH levels significantly increase. Ovulation occurs when the follicle releases the egg. At this point, the follicular phase ends and the luteal phase begins. The left-behind follicular tissue transforms into a corpus luteum, which responds to high LH levels by secreting progesterone. The progesterone inhibits hypothalamic/pituitary production of LH and FSH so that another egg does not mature. The corpus luteum disintegrates and reduces progesterone and estradiol production. These low steroid levels free the hypothalamus and pituitary to repeat the 28-day cycle.

While the egg is maturing, the hormones are preparing the uterus for pregnancy in case the egg is fertilized. Estradiol and progesterone control uterine development. Post-ovulation, estradiol and progesterone stimulate development of the uterine lining by increasing blood flow and glandular development for nutrient attainment. When the corpus luteum degrades and these hormones decline, the uterine wall disintegrates and sloughs off along with blood during menses. Once the follicle develops and estradiol production increases, the uterine lining is developed again, continuing the cycle.

Organisms Communicate and Respond to Stimuli

Behavior is the way in which an organism responds to a stimulus. It can range from turning toward a sound or birds migrating to specific locations at certain times of the year. The sound reaction is an obvious stimulus, but migration is a more complicated stimulus. Studies have shown that day/night circadian clocks, the position of the sun, and even the Earth's magnetic field are interacting stimuli responsible for migratory behavior.

Behavior is not only a response to the environment, but also involves communication and relationships between organisms. Birds communicate to each other in elaborate dances or songs. Humans communicate to each other through language. Fruit fly communication provides an example that demonstrates visual, chemical, tactile, and auditory animal communication, examples of which are outlined below.

- Visual communication: Male fly spies a female and orients his body so that she can see him.
- Chemical communication: Male fly smells a female.
- Tactile communication: Male fly approaches a female and pokes her with his leg.
- Auditory communication: Male fly produces a song via wing vibrations and seduces a female.

Pheromones are chemicals that produce an olfactory stimulus between organisms in a species. Queen bees secrete a pheromone that inhibits ovarian development in her workers and attracts males when she wants to mate, ensuring that her progeny are the only bees to survive.

Innate Behaviors are Inherited

Innate behavior is quite different than behavior based on learning. Learning results in behavior modification derived from experiences. There are some behaviors that are both innate and learned, such as imprinting, which involves a specific period of time during development where an organism is susceptible to learning. The classic example of imprinting focuses on baby geese. There is a period of time not only where the geese offspring mimics its parents and learns species behavior, but also where the parent learns to recognize its offspring. If geese do not bond with their offspring in the first few days, the parents won't care for their infants.

Learning Occurs Through Environmental Interactions

Spatial learning is demonstrated by birds that can return to their nests after hunting, cats that learn where their litter box is, and squirrels that remember where they buried acorns.

More advanced learning involves cognition and problem-solving, and these high-level processes are not just reserved for humans. Rats in mazes and chimpanzee sign language acquisition provide evidence for other mammalian cognition. Even insects have exhibited problem-solving in some studies.

Animal Behaviors are Triggered by Environmental Cues

Behavior is responding to environmental information/stimuli such as phototropism and photoperiodism in plants discussed previously. Animals respond to environmental cues as well. For example, the males in many species of birds have brighter feathers than the females, which is often achieved by the pigments of the fruit in their diet that are deposited into their feathers. The brighter the bird's feathers, the richer and more frequent his diet. Female birds spot these color differences and will choose the male with the brighter colors over the others — ensuring his reproductive success. Her behavior, on the other hand, is in response to his courtship rituals.

Cooperative behavior in ecosystems is also adaptive and contributes to survival. Mutualism, such as a protozoa living in a termite's gut, benefits both organisms; the termite gets aid in digesting wood that it cannot get on its own, and the protozoa gets protection with easily obtained nutrients in exchange.

Practice Questions

1. In aerobic respiration, a hydrogen ion gradient is essential for the production of ATP through oxidative phosphorylation. Which cytoplasmic embedded mitochondria shown below correctly demonstrates the relative concentration of hydrogen ions?

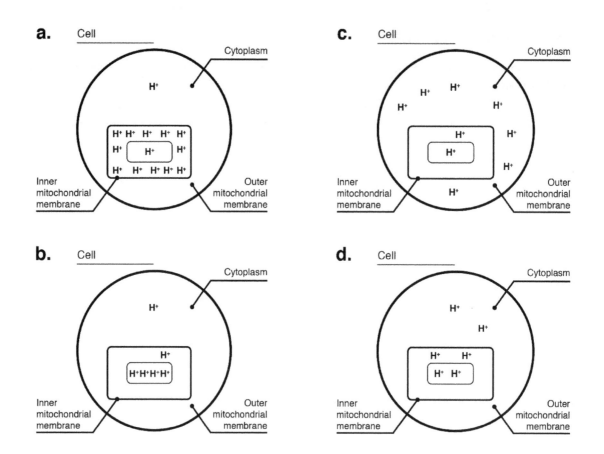

2. Both NADP+ and NAD+ are important for cellular energy conversion. They distribute high-energy electrons to electron transport chains that facilitate the pumping of protons across membranes and couple the action with redox reactions. Which of the following is another similarity between the two molecules?

    a. They both generate ATP by traveling through ATP synthase.

    b. They both carry one proton and a pair of electrons when they are reduced.

    c. They both deliver electrons from either glycolysis or the Krebs cycle to the cristae.

    d. They both are oxidized by electrons energized by the photosystems.

3. Water is a bent molecule due to valence shell electron pair repulsion. The two free electron pairs of oxygen repel, causing the two hydrogen atoms to move away from each other. This is why the ball and stick model looks like a "V." The many unique properties of water are partially due to its molecular shape. The image below shows a soluble ionic compound dissolved in a container of water and the resulting hydration shells. Which statement is true regarding the solution pictured below?

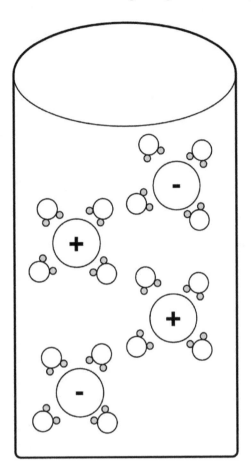

a. Oxygen is more electronegative than hydrogen and pulls the hydrogen electrons closer, resulting in polarity of the solvent.

b. Oxygen is oriented toward the anion because of its partial negative charge due to dipole moments.

c. The water potential increases as more of the ionic compound is added.

d. The hydrogen bonding between the oxygen and its connected hydrogen atoms contribute to a decrease in water potential.

4. The image below shows an experiment that was conducted with the disaccharide maltose. The initial solutions on the left were measured after 30 minutes using a graduated cylinder and a disaccharide indicator. The dotted line represents a membrane that is only permeable to water. Which of the following best explains the results?

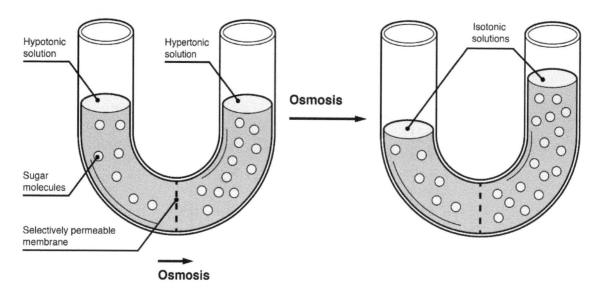

a. The initial solution on the left of the u-tube is hypotonic, so the solvent will move to the right until both sides are isotonic.

b. The initial solution on the right of the u-tube has a greater water potential than the solution on the left, so the solvent will move to the right.

c. The solute moves from the left side of the u-tube to the right until equilibrium is reached.

d. The initial solution on the left of the u-tube has a smaller water potential than the solution on the left, so the solute will move in.

5. What is the solute potential of an open 0.15 M calcium chloride solution at 23 °C?
    a. -3.7
    b. -11.1
    c. -0.29
    d. -0.86

6. Which of the following is true about an endergonic reaction?

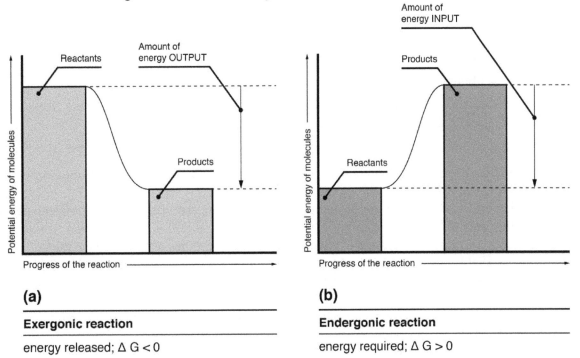

(a)

Exergonic reaction

energy released; $\Delta G < 0$

(b)

Endergonic reaction

energy required; $\Delta G > 0$

    a. The reaction releases energy and decreases entropy.
    b. The reaction absorbs energy and increases entropy.
    c. The reaction releases energy and increases entropy.
    d. The reaction absorbs energy and decreases entropy.

7. The hypothalamus stimulates the pituitary gland with thyroid-stimulating hormone (TSH), and the pituitary stimulates the thyroid gland by producing thyroid-releasing hormone (TRH). Upon activation of the thyroid, hormones T3 and T4, which maintain and stimulate metabolism, are released. TSH also stimulates the thyroid to release calcitonin, which lowers calcium in the blood. Which of the statements below is the most likely regulatory mechanism that fine-tunes metabolism?
    a. As blood calcium levels increase, TSH production also increases.
    b. T_3 and T_4 increase the secretion of calcitonin.
    c. The more TSH released, the more calcium in the blood.
    d. TSH and calcium levels are independent of each other because they have different targets.

8. A study that investigated respiration rates of different organisms recorded the data below. Use the data to answer the question below.

	Hours required for 1 g of the animal to use 10 mL O2
Bird	1.3
Human	2.7
Cat	1.9
Elephant	8.3
Lizard	25.2

Which statement best describes the dramatic difference in the data between the lizard and other organisms?

a. Internal homeostatic mechanisms for temperature regulation are vastly different between lizards and the other animals.

b. Mitochondria structure is different in endotherms and ectotherms, which results in dramatically different metabolic rates.

c. The small size of lizards means that it will have the greatest metabolic rate due to its surface area to volume ratio.

d. Lizards require much less energy to calibrate their temperature because of their habitats.

9. The homology between the circulatory systems of different vertebrates is illustrated in the image below. Which statement best explains the similarities and differences in the physiology between the different organisms?

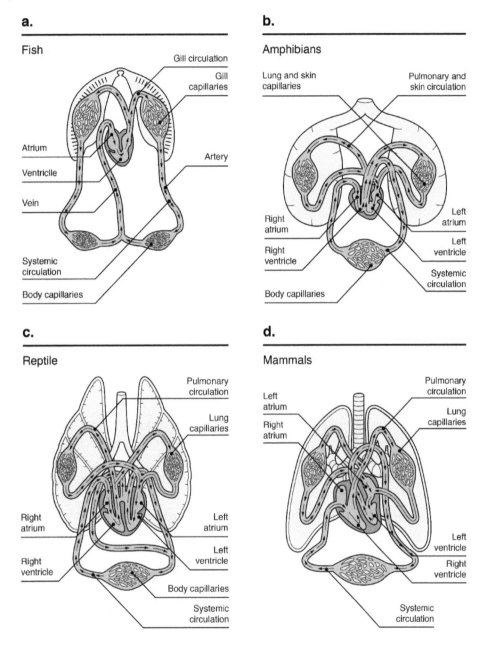

a. The large conservation between amphibians and reptiles suggests that reptiles are the direct descendants of amphibians.

b. Fish are the only group pictured that has gill capillaries, suggesting that there is no common ancestor between fish and the terrestrial groups.

c. The diagrams suggest that reptiles and mammals are more closely related than reptiles and amphibians.

d. The significant difference in the circulatory system between mammals and fish implies that every mammalian organ is more advanced.

10. Two different plants were grown in a lab and their response to light was investigated. Based on the representative qualitative data shown in the diagram, which of these mechanisms would best explain the flowering patterns?

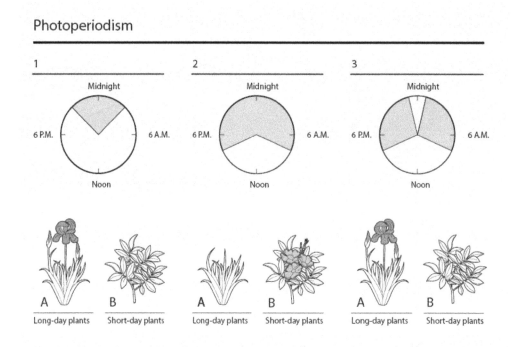

a. Auxin's property of elongating cells exposed to light is responsible for flowering in both plants.
b. Plant B has a selective advantage because it is unusually reproductively active in the winter, which reduces competition.
c. Plant A requires a threshold of sunlight in order to flower. Its blooming is independent of the season.
d. Both plants are dependent on the amount of continuous light exposure in order to flower.

11. The image below shows hormone levels during a 28-day human menstrual cycle. Days 1-14 are when the follicle develops, and day 14 is ovulation and when the egg is released. During days 15-28, the follicle left behind becomes the corpus luteum. This also is the time when the uterine lining vascularizes and develops.

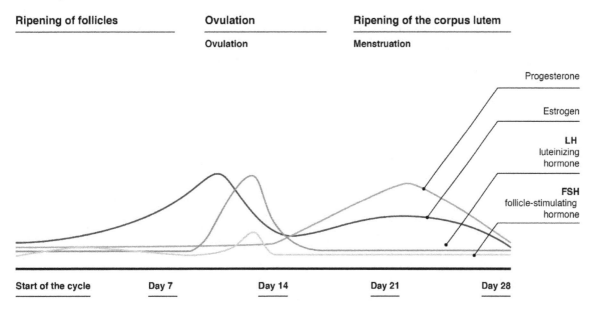

The interrelationship between the hormones is critical in the regulation of the menstrual cycle. Based on the image, which is the most likely and consistently demonstrated conclusion regarding hormone regulation?

    a. FSH is required for progesterone production.

    b. Estrogen is required for LH production.

    c. Progesterone is required for estrogen, LH, and FSH production

    d. FSH and LH require the same negative feedback loop.

12. One reason for women's elaborate control of egg release is that there is a finite number of eggs. Women's eggs are distributed very slowly throughout her fertility window due to the tight hormonal regulation noted in the previous question. How is a woman's energy expenditure benefited by her hormonal regulation?

    a. If the uterine lining were continuously vascularized and ready for egg implantation, metabolic cellular processes would be too costly from an energy perspective.

    b. Menses are preceded by a few days of low energy, where some women become very tired and decrease activity. This energy conservation helps prepare for the next egg release.

    c. The hormonal release simultaneously releases pheromones, which attract men during ovulation and is incidentally women's most fertile time of their cycle.

    d. It is beneficial to release the egg once a month to ensure only one egg is fertilized so that pregnancy does not drain the body of too many resources.

13. The human immune system has an arsenal of white blood cells. The following image shows the sequence of events in the humoral immune response. How would this process be affected if the individual had been vaccinated against the pathogen prior to infection?

a. Antibodies would be freely circulating and cause a more efficient secondary immune response.

b. T cells would be circulating and begin to produce antibodies much quicker due to the vaccination.

c. Existing B cells will quickly differentiate into plasma cells upon cytokine stimulation, causing a more robust and faster response.

d. Phagocytic lymphocytes are already presenting MHC class II molecules that activate the immune response.

Clonal Selection and Ensuing Events

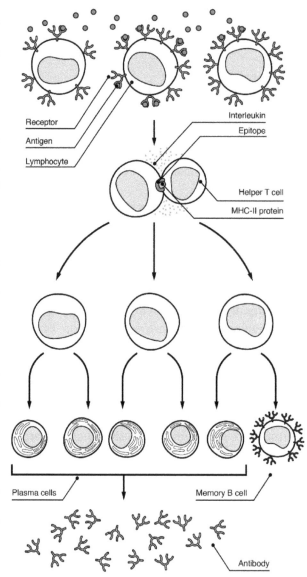

1

Antigen recognition

Immunocompetent B cells exposed to antigen. Antigen binds only to B cells with complementary receptors.

2

Antigen presentation

B cell internalizes antigen and displays processed epitope. Helper T cell binds to B cell and secretes interleukin.

3

Clonal selection

The Interleukin stimulates the B cell to make it divide repeatedly and form a clone.

4

Differentiation

Some cells of the clone become memory B cells. Most differentiate into plasma cells.

5

Attack

The plasma cells synthesize and secrete antibodies. These antibodies then employ various means to render antigens harmless.

14. Blood type is a trait determined by multiple alleles, and two of them are co-dominant: IA codes for A blood and IB codes for B blood. i codes for O blood and is recessive to both. If an A heterozygote individual and an O individual have a child, what is the probably that the child will have A blood?
    a. 25%
    b. 50%
    c. 75%
    d. 100%

15. Microvilli in the small intestine serve an important function by increasing nutrient absorption. The three conditions in the table all affect blood sugar circulation in different ways, which are all indirectly related to absorption. The more efficient the absorption, the more sugar enters the bloodstream. Type I diabetes involves little to no insulin production, which causes very high blood sugar if not treated with exogenous insulin. People with celiac disease have trouble digesting gluten. Hypoglycemia results in very low blood sugar.

Microvilli Size in Different Groups of Individuals		
	Average Length (μm)	Average Width and Height (μm)
Unaffected	4.7	0.8
Type I diabetes	3.3	1.7
Celiac	1.1	1.7
Hypoglycemia	5.0	0.5

Which statement is a reasonable conclusion given the data recorded? The sample size was very small, and although these numbers do not necessarily reflect that of the whole population, assume that they do.
    a. The individuals with celiac disease have the smallest surface area to volume ratio, which will be beneficial because it will make nutrient absorption more efficient.
    b. The individuals with hypoglycemia have a larger surface area to volume ratio, which will be beneficial because it will make nutrient absorption more efficient.
    c. The unaffected individuals have the largest surface area to volume ratio, which will be beneficial because it will make nutrient absorption more efficient.
    d. The unaffected individuals have the smallest surface area to volume ratio, which will be beneficial because it will make nutrient absorption more efficient.

16. A characteristic of life is that organisms are able to react and respond to their environment. The choices below are proposed physiological regulatory responses of various organisms. Which one actually occurs?
    a. Fungi rely on phototropism to aid in obtaining resources.
    b. Humans regulate temperature solely by sweating and shivering.
    c. Plants maintain circadian rhythms, such as timing when their stomata open.
    d. Protists use the hormone endocrine-driven feedback loops to regulate processes.

17. What is the term used for the set of metabolic reactions that convert chemical bonds to energy in the form of ATP?
    a. Photosynthesis
    b. Reproduction
    c. Active transport
    d. Cellular respiration

18. To answer the following question, refer to the diagram below:

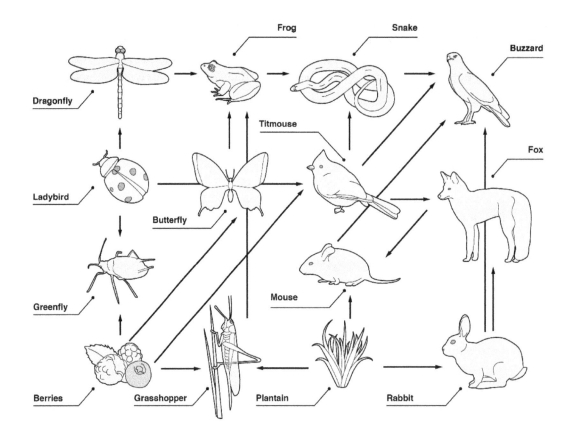

Assume that the snake population has been wiped out by a disease that is only transmutable between snakes. Which of the following would most likely be true as a result?
    a. The dragonfly population would decrease.
    b. The grasshopper population would increase.
    c. The fox population would decrease.
    d. The buzzard population would increase.

19. Amphibian development is different than mammalian development in that it involves a larval stage and metamorphosis. Unlike the caterpillar/butterfly metamorphosis cycle, the stages of larval development in the tadpole/frog conversion are readily observable. Identify the most likely mechanism for the exchange of the tail for legs in frog development.

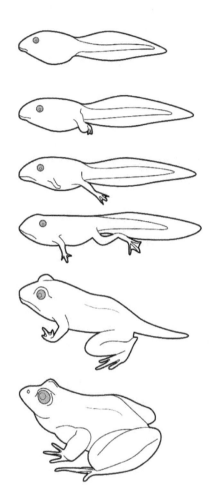

a. The tails are destroyed via necrosis, causing them to grow legs in order to replace the tail.
b. While the organs are developing, homeotic gene factors induce formation of the notochord and primitive brain, which signals the posterior region to die.
c. Tail cells undergo apoptosis, which slowly degrades tail tissue while cell determinants initiate leg development.
d. A feedback loop in the frog evaluates environmental conditions and sends hormones to initiate cell death in the tail when water is scarce.

20. What are the molecular "energy" investments necessary for every G3P molecule produced by the Calvin cycle?
a. 9 ATP molecules and 6 NADPH molecules
b. 9 ATP molecules and 6 NADP⁺ molecules
c. 6 ATP molecules and 9 NADP⁺ molecules
d. 6 ATP molecules and 9 NADPH molecules

Answer Explanations

1. A: In the process of oxidative phosphorylation, it is critical that there is a concentration gradient where hydrogen ions are at higher levels in the inner membrane space and are at lower levels in the matrix. The mitochondria use energy from the high-energy electrons that NADH and $FADH_2$ are carrying. The energy is used to pump protons across the cristae from the matrix to the inner membrane space. This is significant because it maintains the gradient that is responsible for the facilitated diffusion of the hydrogen ion across the cristae through ATP synthase. ATP synthase spins when the proton goes back through, providing the kinetic energy for the endergonic reaction that creates ATP.

ADP + P = ATP

2. B: $NADP^+$ and NAD^+ are very similar molecules in that they carry a proton and steal a pair of high-energy electrons when they are reduced to NADPH and NADH. The major difference is that $NADP^+$ is the electron carrier in photosynthesis and NAD^+ is the electron carrier in respiration.

$NADP^+$, not NAD^+, takes electrons from the electrons excited by Photosystem I, making Choice *D* incorrect, not to mention that they are not oxidized. NAD^+, not $NADP^+$, is responsible for reducing high-energy glucose derivatives in glycolysis and in the Krebs cycle, making choice C incorrect.

The electrons of reduced $NADP^+$ are invested in sugars in the Calvin cycle, and the electrons from reduced NAD^+ are delivered to an electron transport chain in the cristae. Neither molecule travels through ATP synthase, making Choice *A* incorrect.

3. A: Oxygen is far more electronegative than hydrogen and has more "proton pull," meaning that it has a larger positively charged nucleus. The two hydrogen atoms' significantly smaller positively charged nucleus does not attract their electrons very strongly, which allows the electrons to have more freedom and mobility. This is why the oxygen nucleus is able to pull in the hydrogen electrons, which results in the hydrogen atoms' partial positive charge and the oxygen atom partial negative charge, creating a polar molecule. This "negative" oxygen will be attracted to the positive cation, not the negative anion, making Choice *B* incorrect. Note that there is no official "bond" in hydration shells. What is occurring is an intermolecular attraction due to hydrogen bonding the attractions. Choice *D* states that hydrogen bonding is intramolecular, and is therefore incorrect. Regarding water potential, it decreases as more solute is added, which is why Choice *C* is incorrect.

4. A: Hypotonic means that a solution is less concentrated than the one that it is being compared to and therefore, has a lower concentration of solute and a higher concentration of water. The solution on the left is hypotonic, and due to the principles of passive transport, water will move from high concentration to low concentration and will therefore, move from the left side to the right side until both sides are isotonic, making Choice *A* correct. Solute concentrations are unchanged between the two sides after equilibrium is reached, illustrating that the membrane is semipermeable and that the disaccharide maltose is too large to pass through, throwing out Choices *C* and *D* that refer to solute movement. Water potential decreases as solute increases, and because the solution on the right side of the u-tube has more solute, it therefore has a lower water potential. This makes Choice *B* incorrect.

5. B: Solute potential is determined by the following equation.

$$\Psi_s = -iCRT$$

The variables represent the following:

- First, i is the ionization constant and accounts for the fact that ionic compounds dissociate. Calcium chloride ($CaCl_2$) will dissociate into Ca^{+2} and $2Cl^{-1}$, meaning that for every calcium chloride molecule, there will be 3 ions circulating. Remember that solute potential is the pressure needed (provided by solutes) that will prevent the inward flow of water. Solute potential decreases as solute increases, similar to when obstacles are added to an obstacle course, it makes movement more difficult. In this particular equation, i equals 3.

- C represents the molar concentration, in this case 0.15 $CaCl_2$. The higher the concentration, the lower the solute potential.

- T represents temperature and it needs to be in Kelvin (°C + 273). In this equation, it will be equal to 23 + 273, which equals 296.

- R is the pressure constant, which is .0831 L x bar/(mole x K).

Substitute to find the solution

Ψ_s = -3(0.15 mole L) x .0831 L bar/mole x K x (296K) = -11.1 bar

The other answer choices are calculations using either the wrong ionization constant or lacking the Kelvin conversion.

6. D: An endergonic reaction absorbs energy to make bigger things from smaller things, resulting in an increase in order, or a decrease in entropy, because the molecules become condensed into a more rigid form. Choice *A* is incorrect because it does not release energy, it absorbs it. Choice *B* is incorrect because it does not increase entropy. Choice *C* is incorrect because it does not release energy and increase entropy—these are the requirements of an exergonic reaction.

7. A: This is a negative feedback loop that fine-tunes hormone production to maintain homeostasis. If calcium levels are too high, the hypothalamus detects it and stimulates TSH to lower the levels. If calcium levels are too low, the hypothalamus detects it and inhibits TSH production. Choice *D* is incorrect because TSH and calcium cooperate in the feedback loop as shown in the diagram below. Choice *C* is incorrect because TSH ultimately results in decreased calcium levels. Choice *B* is also false because calcitonin is regulated by TRH, not T_3 and T_4.

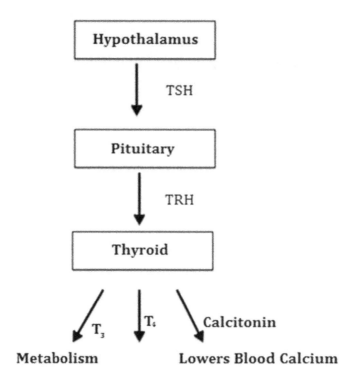

8. A: The data table is referring to rate of respiration. The more time required to use oxygen indicates a lower rate of respiration. The quicker the animal uses up oxygen, the more aerobic respiration is taking place. Ectothermic and endothermic animals are different, since endotherms have to regulate their own internal temperature, which requires more ATP because it is controlled by an elaborate feedback loop. More ATP requires higher levels of respiration. Choice *C* is attractive because it is true that smaller organisms respire faster due to their larger surface area to volume ratio. However, Choice *C* is incorrect for two reasons. First, not all lizards are smaller than the other organisms. Secondly, the dramatic nature of the difference is due to the lizards' simpler mode of temperature regulation. Choice *D* is attractive because lizards bask in sunny habitats to help maintain ideal temperatures, but they lack a feedback loop that calibrates temperature, so it is incorrect. Choice *B* is incorrect because mitochondrial structure is the same among all eukaryotes—it does not change.

9. C: The conservation of the four-chambered heart provides the justification for concluding that the greater relatedness is between mammals and reptiles. As this is somewhat subjective due to qualitative interpretation, it is still the best choice. Choice *A* is incorrect because organisms are not necessarily direct descendants of more primitive versions of existing organisms; they could have branched from along different points in the phylogenetic tree. Choice *B* is incorrect because all organisms share a common ancestor. Choice *D* is incorrect because it is impossible to extrapolate broad conclusions based on qualitative data regarding one body system. Fish do, in fact, have very complicated organs such as gills that are highly differentiated and complex.

10. B: Photoperiodism is the phenomenon where dark exposure determines flowering, and therefore Choices *C* and *D* are wrong. Choice *B* is correct because plant B is dependent on the amount of darkness—not on the amount of light. The interrupted darkness in scenario three prevents flowering of the plant. Flowering in the winter, or when the amount of light is limited, is unusual because usually springtime is when flowers make their appearance. This could be considered an adaptation because it reduces competition for pollinators, since other flowers are not present to compete, ensuring the plant passes on its genes. Choice *A* is untrue because auxin is responsible for stem elongation and is not related to flowering.

11. D: This data suggests that FSH and LH are both regulated by the same mechanism because they spike at the same time. Choice *A* is incorrect because while right after the FSH spike, progesterone levels increase, FSH immediately decreases and progesterone is unaffected. Choice *B* is incorrect because estrogen has no effect on LH in the luteal phase. Choice *C* is not the best choice because in the follicular phase, the levels of progesterone, LH and FSH are all low.

12. A: Organisms are constantly balancing energy expenditure to do three things: maintain homeostasis, grow, and reproduce. The uterus is only necessary during pregnancy. Keeping it ready throughout a woman's entire menstrual cycle would be, using a baseball analogy, like standing at home plate in batting position when there is not a game going on. Maintaining the uterine lining when it isn't necessary is a waste of energy that could be used in other cellular processes. Choice *B* is incorrect because conserving energy for an upcoming event is usually associated with seasonal events such as hibernation, where energy is stored in fats before the body goes into "sleep mode." Choice *C* is incorrect because it is referring to reproductive animal behavior and not energy conservation and expenditure. Choice *D* is actually an attractive answer. If eggs were always available like sperm, a woman could theoretically have hundreds of implanted eggs, which would suck every resource and every bit of energy out of her until she died, and then the embryos couldn't live because they would starve. So, it is beneficial to release the egg monthly. The nature of the question, however, makes this choice incorrect because it is specifically asking about the hormonal regulation. To have all eggs out there at once would make the cycle hormone-independent.

13. C: During vaccination, the immune response is initiated. This is called the primary, or first, immune response. Any subsequent response is called the secondary immune response because the "blueprints" to make the antibodies are circulating actively in memory B cells. The memory B cells are prepared for a "second" exposure because they can recognize the antigen. Choice *C* is the right answer because the memory of the vaccination stimulates B cells to proliferate. Choice *A* is incorrect because memory B cells, not antibodies, are circulating. Choice *B* is incorrect because B cells, not T cells, produce antibodies. Choice *D* is incorrect because phagocytes do not present antigens until after ingestion of a pathogen.

14. B: 50%. According to the Punnett square, the child has a 2 out of 4 chance of having A-type blood, since the dominant allele I^A is present in two of the four possible offspring. The O-type blood allele is masked by the A-type blood allele since it is recessive.

I^A i	ii
I^A i	ii

15. B: The question is based on absorption efficiency and surface area to volume ratio. The larger the ratio, the better the cell will be at absorption because there is more surface area for absorption to occur. The first step is to calculate the surface area to volume ratios of the different groups. This can be done by likening the microvilli structure to a rectangular prism or cylinder, which will not give exact answers, but will serve as a model that will reflect the trend. The following calculations use a rectangular prism model for calculations.

Surface area of a rectangular prism minus the side where it is attached to the small intestine is:

length x width x 4 + the outer face, which is width x height

The volume of the rectangular prism is:

length x width x height.

To find the surface area to volume ratio, divide the surface area by the volume. The highest surface area to volume ratio is the most efficient microvilli, meaning it is the one absorbing the most nutrients.

Group	Surface Area	Volume	Ratio
	L x W x 4 + W x H	L x W x H	
Unaffected	15.7	3.0	5:1
Type I diabetes	25.3	9.5	3:1
Celiac	10.4	3.2	3:1
Hypoglycemia	10.3	1.3	8:1

Choices *C* and *D* are wrong because the unaffected group does not have the largest or smallest surface area to volume ratio. Choice *A* is wrong because while the celiac group does have the smallest surface area to volume ratio, that makes it the least efficient, not the most. Choice *B* is the best choice because a large surface area to volume ratio results in more efficient transport.

16. C: Fungi are not photosynthetic, making Choice *A* incorrect. Choice *B* is incorrect because temperature regulation is more complicated than that. For example, induction of fever as part of the immune system's response to pathogens is under the umbrella of temperature regulation. Choice *D* is incorrect because protists are single-celled organisms for the most part, and the multi-cellular ones are not differentiated enough to have organs; therefore, they do not have an endocrine system.

17. D: Cellular respiration is the term used for the set of metabolic reactions that convert chemical bonds to energy in the form of ATP. All respiration starts with glycolysis in the cytoplasm, and in the presence of oxygen, the process will continue to the mitochondria. In a series of oxidation/reduction reactions, primarily glucose will be broken down so that the energy contained within its bonds can be transferred to the smaller ATP molecules. It's like having a $100 bill (glucose) as opposed to having one hundred $1 bills. This is beneficial to the organism because it allows energy to be distributed throughout the cell very easily in smaller packets of energy.

18. A: If the snake population were to disappear, the organisms that it hunted would increase, and the food that those organisms ate would decrease. Without the snake, that means there are more frogs, which in turn means there are more frogs to eat more dragonflies, so the dragonfly population would decrease. Choice *B* is incorrect, since an increase in frogs means more frogs would eat more grasshoppers, so grasshoppers would decrease as opposed to increase. Choice *C* is false because fewer snakes means more titmouses, which provide more food for the foxes and results in an increase in the fox population, not a decrease.

19. C: Tadpole tail degradation is a classic example of apoptosis, or programmed cell death, which is an important process in morphogenesis of organisms. Apoptosis explains why some animals have webbed feet and some don't. Choice *A* is incorrect because necrosis is cell death induced by injury, and leg formation in tadpoles is not caused by injury. Choice *B* is wrong because apoptosis is not a nervous system-controlled event; it is either activated by an individual cell's internal signals or neighboring cell signals. Choice *D* is incorrect because apoptosis is not controlled by hormones. Choice *D* is incorrect because fewer snakes means less food for the buzzards to eat, resulting in a decrease in population.

20. A: The Calvin cycle is dependent upon the ATP and NADPH produced by the light reactions. Nine ATP molecules and six NADPH molecules are invested into the Calvin cycle for every one molecule of glyceraldehyde 3-phosphate (G3P) produced. In the endergonic reduction reaction, NADPH uses energy from ATP to add a hydrogen to each molecule of 3-phosphoglycerate. This converts the six molecules of 3-PGA into the 3-carbon sugar G3P. ATP supplies energy by donating a phosphate group (Pi) (becoming ADP), and NADPH loses a hydrogen to become $NADP^+$.

Living Systems Storage, Retrieval, Transmittal, and Response to Information

The Source of Heritable Information

DNA was not always believed to be the cell's genetic material. Significant scientific contributions included:

<u>Griffith, 1920s</u>

Experimental Design & Results
- Infected mice with harmful and benign bacteria.
- If harmful bacteria were killed and injected, no effect
- If harmful bacteria were killed and mixed with living benign bacteria, some part of the harmful bacteria transformed the harmless bacteria

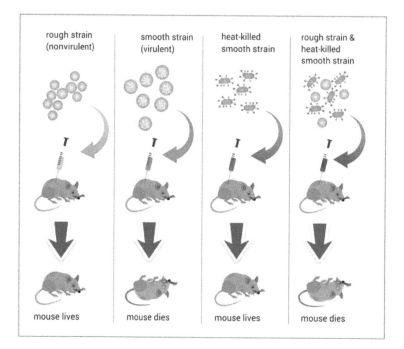

Discovery & Significance
- Something in a cell is capable of passing instructions
- Coined the term "transformation" or the process when a cell picks up foreign DNA and then changes phenotype

Avery, 1930s & 1940s

Experimental Design & Results
- Did the same experiment except he used a substance to deactivate either DNA, RNA, or protein

- He found only when DNA was active did the harmless bacteria transform the cell, indicating that DNA, not RNA or protein, is the genetic material.

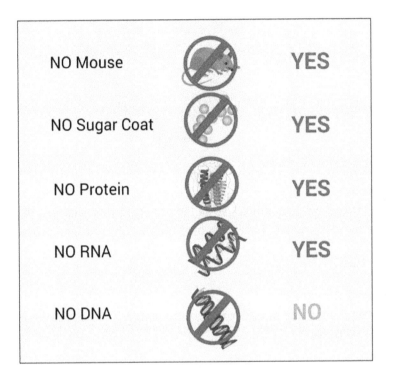

Discovery & Significance
- Skeptics dismissed the results due to the belief that bacteria and eukaryotes have different processes.

Hershey & Chase, 1950s

Experimental Design & Results
- Different radiolabels were used on proteins and DNA in phages that infected bacterial cells.
- It was observed that the DNA label was inside the bacteria while the protein label was in the supernatant.

Discovery & Significance
- DNA is the component that enters cells, therefore DNA is the genetic material.

Rosalind Franklin and Watson & Crick 1940s & 1950s

Experimental Design & Results
- Used X-ray crystallography to photograph DNA
- Used Franklin's data to finalize and publish the double-helix model of DNA

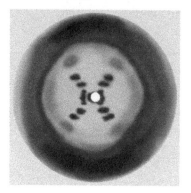

Discovery & Significance
- Discovered that the coiled, double stranded molecule was arranged in an anti-parallel manner
- DNA had a sugar-phosphate backbone held together by nitrogen bases

Meselson & Stahl, 1950s

Experimental Design & Results
- Radiolabelled DNA with a heavy isotope, then placed it in a solution where it replicated with a lighter labeled isotope

- If DNA replication is conservative, then the original heavy DNA would still be present (not observed).

- If DNA replication is dispersive, there would only be one thick medium band (not observed).

- Semi-conservative replication would result in a thin medium band and more and more light bands since the DNA was only replicating with the light isotope (observed).

Discovery & Significance
- DNA replication is semi-conservative, meaning that when the strands are separated, new DNA is built from each parent strand. Each new DNA molecule has one old strand and one new strand.

Retroviruses Use an Alternate Flow for Genetic information

Retroviruses, such as HIV, are a type of RNA virus that must enter the host cell in RNA form and then be converted to DNA form. This process, which takes place in the opposite direction of typical information flow in a cell, is called reverse transcription and is carried out by an enzyme coded for by the viral genome called reverse transcriptase. As soon as the retrovirus enters the host cell, the RNA is transcribed. This produces reverse transcriptase, which in turn, produces a copy of the viral genome's DNA. At some point, depending on the specific retrovirus, the host chromosome excises the viral DNA and the DNA becomes active and produces new viruses.

Structure and Function of DNA

Deoxyribonucleic acid (DNA) is life's instruction manual. It is double stranded and directional, meaning it can only be transcribed and replicated from the *3' end* to the *5' end*.

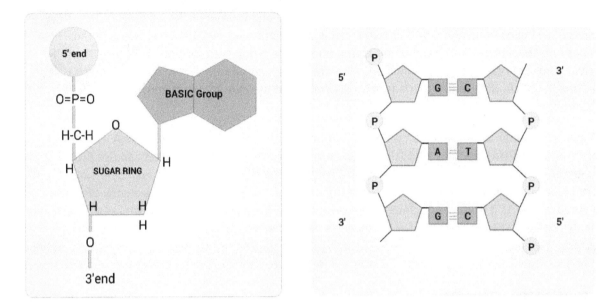

DNA as a Monomer	DNA as a Polymer
A nucleotide is composed of a five-carbon sugar with a Phosphate Group off of the 5th carbon and a Nitrogen Base off of the 1st carbon. DNA and RNA are different because DNA contains deoxyribose sugar while RNA contains ribose sugar. Also, the nitrogen base thymine in DNA is replaced by uracil in RNA.	The two strands are antiparallel, meaning they are read in opposite directions. The bases guanine and cytosine are complementary and held together by three hydrogen bonds, and the bases adenine and thymine are complementary and held together by two hydrogen bonds. Weak hydrogen bonding between bases allows DNA to be opened easily for transcription and replication.

Nitrogen bases come in two varieties:

- *One-Ringed Pyrimidines*: Cytosine and Thymine (DNA)/Uracil (RNA)
- *Two-Ringed Purines*: Adenine and Guanine

Purines and Pyrimidines

The five bases in DNA and RNA can be categorized as either pyrimidine or purine according to their structure. The pyrimidine bases include cytosine, thymine, and uracil. They are six-sided and have a single ring shape. The purine bases are adenine and guanine, which consist of two attached rings. One ring has five sides and the other has six. When combined with a sugar, any of the five bases become nucleosides. Nucleosides formed from purine bases end in "osine" and those formed from pyrimidine bases end in "idine." Adenosine and thymidine are examples of nucleosides. Bases are the most basic components, followed by nucleosides, nucleotides, and then DNA or RNA.

RNA

RNA is short for ribonucleic acid, which is a type of molecule that consists of a long chain (polymer) of nucleotide units. RNA and DNA differ in terms of structure and function. RNA has a different sugar than DNA. It has ribose rather than deoxyribose sugar. The RNA nitrogenous bases are adenine (A), guanine (G), cytosine (C), and uracil (U). Uracil is found only in RNA, and thymine is found only in DNA. RNA consists of a single strand and DNA has two strands. If straightened out, DNA has two side rails. RNA only has one "backbone," or strand of sugar and phosphate group components. RNA uses the fully hydroxylated sugar pentose, which includes an extra oxygen compared to deoxyribose, which is the sugar used by DNA. RNA supports the functions carried out by DNA. It aids in gene expression, replication, and transportation.

RNA acts as a helper to DNA and carries out a number of other functions. Types of RNA include ribosomal RNA (rRNA), transfer RNA (tRNA), and messenger RNA (mRNA). Viruses can use RNA to carry their genetic material to DNA. Ribosomal RNA is not believed to have changed much over time. For this reason, it can be used to study relationships in organisms. Messenger RNA carries a copy of a strand of DNA and transports it from the nucleus to the cytoplasm. DNA unwinds itself and serves as a template while RNA is being assembled. The DNA molecules are copied to RNA. Translation is the process whereby ribosomes use transcribed RNA to put together the needed protein. Transfer RNA is a molecule that helps in the translation process, and is found in the cytoplasm. Ribosomal RNA is in the ribosomes.

Transcription and Translation

Transcription is the process by which a segment of DNA is copied onto a working blueprint called RNA. Each gene has a special region called a promoter that guides the beginning of the transcription process. RNA polymerase unwinds the DNA at the promoter of the needed gene. After the DNA is unwound, one strand or template is copied by the RNA polymerase by adding the complementary nucleotides, G with C, C with G, T with A, and A with U. Then, the sugar phosphate backbone forms with the aid of RNA polymerase. Finally, the hydrogen bonds joining the strands of DNA and RNA together are broken. This forms a single strand of messenger RNA or mRNA.

Codons are groups of three nucleotides on the messenger RNA, and can be visualized as three rungs of a ladder. A codon has the code for a single amino acid. There are 64 codons but 20 amino acids. More than one combination, or triplet, can be used to synthesize the necessary amino acids. For example, AAA (adenine-adenine-adenine) or AAG (adenine-adenine-guanine) can serve as codons for lysine. These groups of three occur in strings, and might be thought of as frames. For example, AAAUCUUCGU, if read in groups of three from the beginning, would be AAA, UCU, UCG, which are codons for lysine, serine, and serine, respectively. If the same sequence was read in groups of three starting from the second position, the groups would be AAU (asparagine), CUU (proline), and so on. The resulting amino acids would be completely different. For this reason, there are start and stop codons that indicate the beginning and ending of a sequence (or frame). AUG (methionine) is the start codon. UAA, UGA, and UAG, also known as ocher, opal, and amber, respectively, are stop codons.

Ribosomes synthesize proteins from mRNA in a process called translation. Sequences of three amino acids called codons make up the strand of mRNA. Each codon codes for a specific amino acid. The ribosome is composed of two subunits, a larger subunit and a smaller subunit, which are composed of ribosomal RNA (rRNA). The smaller subunit of RNA attaches to the mRNA near the cap. The smaller subunit slides along the mRNA until it reaches the first codon. Then, the larger subunit clamps onto the smaller subunit of the ribosome. Transfer RNA (tRNA) has codons complementary to the mRNA codons.

The tRNA molecules attach at the site of translation. Amino acids are joined together by peptide bonds. The ribosome moves along the mRNA strand repeating this process until the protein is complete. Proteins are polymers of amino acids joined by peptide bonds.

Genes and Biotechnology

The table below lists some biotechnology tools that scientists use to study molecular genetics:

Step/Technology	Explanation
Polymerase Chain Reaction (PCR)	Primers are developed upstream and downstream of the gene of interest (engineered with deliberate restriction enzyme recognized sequences). DNA polymerase is added to the mix. There is a three-step cycle consisting of heat, cool, and warm phases which is then repeated many times. Heat separates and denatures DNA strands. Cool is where the primers anneal. Warm is when DNA polymerase elongates and synthesizes DNA.
Restriction Enzyme Digest	Restriction enzymes are a primitive bacterial defense designed to cut specific sequences of DNA. Scientists utilize them to cut open plasmids and insert their gene of interest. The same restriction enzyme is also used to cut the ends of the gene of interest. Ligase is used to glue the gene of interest into the plasmid (complementary overhanging ends anneal). *Bacterial plasmid usually carries an antibiotic resistant gene.*
Transformation	Transform bacteria with recombinant DNA. Grow bacteria on an agar plate treated with antibiotic. Only surviving clones contain the gene of interest.
Gel Electrophoresis	Separates DNA based on sequence length. DNA is negatively charged so, when placed in a gel with an electric field applied, it travels to the positive electrode and smaller segments travel faster.

The Cell Cycle has Checkpoints

Mitosis has several checkpoints. If a cell exits the cell cycle, it enters a phase called G0 that consists of non-dividing cells like neurons. G1 checkpoints prepare and commit cells to entering the cell cycle. S phase proofreads and corrects mistakes. There is also a G2 checkpoint: as a cell progresses through G1,

S, and G2, the cyclin protein accumulates. When it becomes abundant, it binds with a cyclin dependent kinase (cdk) to form *Maturation Promoting Factor (MPF)*, an activating protein complex that facilitates mitosis. As mitosis is completed, the cyclin is degraded, the MPF complex disassembles, and the cell cycle begins once again.

Cell Cycle in Eukaryotes: Interphase

The eukaryotic cell cycle involves four phases: growth (G_1), replication of DNA (S), preparation for cell division (G_2), and cell division (M). The three phases preceding cell division are collectively called *interphase*.

S Phase: DNA replication

A cell must duplicate its DNA prior to cell division so that each daughter cell receives the full set of genetic instructions. To do so, cells use many enzymes. Helicase binds to an origin of replication, separates the two DNA strands, and forms a replication bubble. It then travels along the double-stranded DNA, unzipping it. Topoisomerase is an enzyme downstream of helicase that prevents supercoiling of the DNA. It does this by snipping the DNA, swiveling around it, and then pasting the strands back together. Single-stranded binding proteins hold the DNA strands apart at the replication fork.

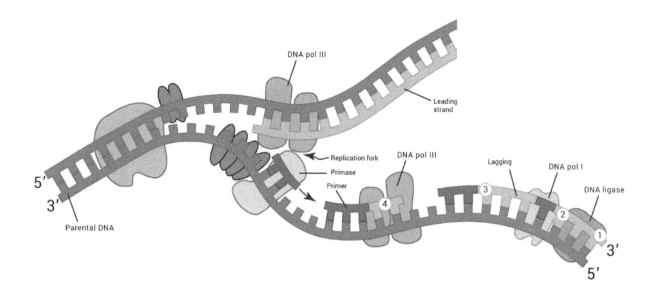

The two DNA strands are copied differently.

Leading Strand – Continuous	Lagging Strand – Interrupted
In this strand, DNA is replicated continuously as the helicase travels. DNA polymerase III is the elongating enzyme that copies the DNA in a complementary fashion. Before DNA polymerase III can attach to the DNA and begin, an enzyme called *primase* must insert a small RNA primer for the polymerase to bind to because DNA polymerase III is unable to bind directly to DNA.	This synthesis is similar to the leading strand in that primase deposits a primer for DNA polymerase III binding. However, because DNA cannot be created in the 3' to 5' direction, DNA polymerase III replicates DNA in small 5' to 3' chunks called *Okazaki fragments*. The pieces are eventually connected after DNA polymerase I replaces the RNA primer with DNA. An enzyme called *ligase* glues the small fragments together.

The cell has several levels to correct mistakes during replication:

- DNA polymerase can proofread as it goes.
- Other enzymes can mismatch repair if polymerase fails to recognize mutations.
- Nucleotide excision repair is a higher-tier corrective mechanism. It involves an enzyme called *nuclease* snipping out the mutation, followed by ligase gluing in the correct sequence.

Telomeres are non-coding, repeating sequences of DNA at the ends of each chromosome. The ends of the chromosome lack a 3' end for DNA polymerase to attach to and replace the RNA primer. As a result, DNA polymerase is unable to bind and, at each division, a little bit of the ends of each chromosome are un-replicated and lost. When the telomeres shorten to nonexistence (at *senescence*), it's a signal to the cell to end its cell cycling days and go through the process of *apoptosis* or intentional cell death. This could have evolved as a natural defense against cancer since older cells, which are more prone to mutations, are killed off before they become dangerous.

Germ cells have the enzyme *telomerase* that elongates telomeres. This is important for reproductive purposes so that *gametes* (egg and sperm cells) can continue to have full DNA instructions.

M Phase

Following DNA synthesis is the G2 phase, where the cell assembles the machinery necessary for cell division. Cell division then occurs in several stages:

PHASE	PHASE EVENTS	ANIMAL CELL DIAGRAM	PLANT CELL DIAGRAM
Prophase	Nucleus disappears and DNA condenses into chromosomes. DNA is already wrapped around histone proteins, and it continues to supercoil until it looks like the letter X. Sister chromatids on either side of the X are identical.		
Metaphase	Chromosomes line up in the cell's center. Kinetochore microtubules extend from animal centrosomes that contain centrioles (organizing centers) on either side of the cell and attach to the centromeres (repeating sequences in the middle of the chromosomes). Nonkinetochore proteins elongate animal cells. This massive protein orchestra is collectively called the spindle apparatus. Plants lack centrioles but have microtubule organizing centers.		
Anaphase	Kinetochore microtubules shorten/are pulled in and sister chromatids are separated and move to each daughter cell.		
Telophase and Cytokinesis	Nuclei reform and chromosomes decondense within them.\n\n**Animals:** actin and myosin microfilaments pinch off the cytoplasm at the cleavage furrow to form two new cells\n**Plants:** cell plate (new cell wall) forms between daughter cells and extends to divide into two new cells		

Mitosis occurs for many reasons:

- Development and growth of an organism
- Differentiation and specialization in multicellular organisms
- Replacement of cells with damage or rapid turnover

Cancer cells are the result of inappropriate cell division and occur when cells are unresponsive to checkpoint regulation and growth factor signals. Normal cells are density dependent and anchorage dependent, but cancer cells lose these properties.

Meiosis Ensures Genetic Diversity in Sexually-Reproducing Organisms

Meiosis is similar to mitosis because it involves cellular division. However, while mitosis involves the division of somatic (body) cells, meiosis is specifically the production of gametes (egg and sperm). In mitosis, one parent cell splits once into two genetically identical and diploid daughter cells, while in meiosis, one germ cell splits twice into four genetically different, haploid daughter cells. The two divisions of meiosis (*meiosis I* and *meiosis II*) are critical because, when a sperm fertilizes an egg to create the first cell of a new organism (*zygote*), the zygote must have two sets of chromosomes—not four—to be viable.

The image below shows the steps of meiosis I. These steps look the same as those of mitosis except that the homologous chromosomes line up in the middle of meiosis I. This formation of homologous chromosome pairs is called a *tetrad*.

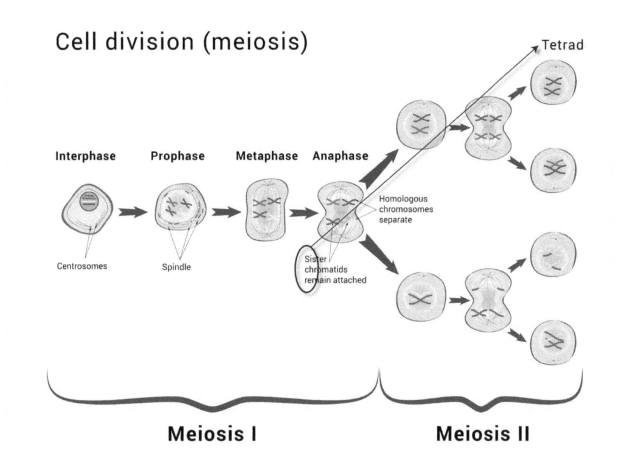

In mitosis, homologous chromosomes line up single file in metaphase. On the other hand, tetrads are paired chromosomes and, in meiosis I, go through a process called *crossing over/recombination* where they exchange DNA. The tetrad is called the *synaptonemal complex*, and a *chiasmata* is where the crossing over occurs (there are 1-3 crossing over events per tetrad). Crossing over makes gametes unique because genetic recombination events are random and unpredictable.

After meiosis I, the homologous chromosomes separate, and the two daughter cells are *haploid*. There is no interphase between meiosis I and meiosis II, so the DNA is not replicated. In anaphase II of meiosis, sister chromatids separate and, by the end, there are four unique, haploid daughter cells.

Although homologous chromosomes code for the same genes, there is diversity. This is because each gene has two or more *alleles*. Alleles are different forms of genes, and the *phenotype* is dependent on the *genotype*.

A process called crossing over occurs, which makes the daughter cells genetically different. If chromosomes didn't cross over and rearrange genes, siblings could be identical clones. There would be no genetic variation, which is a critical factor in the evolution of organisms.

Chromosomal Basis of Inheritance

Tenets of Mendelian genetics:

- *The law of dominance*: Dominant alleles trump recessive alleles in phenotype (the exceptions are non-Mendelian traits)

- *The law of segregation*: Alleles for each trait are separated into gametes. One allele comes from each parent, giving the offspring two copies of each allele.

- *The law of independent assortment*: Tetrads line up in metaphase I independently of other chromosomes. Each of the 23 homologues has a 50/50 chance of being on either side.

In simple Mendelian genetics, an individual can have three different genotypes. This is shown in the table below regarding the trait of flower color:

Genotype	Referred to as	Corresponding Phenotype
PP	Homozygous dominant	Purple flowers
Pp	Heterozygous	Purple flowers
pp	Homozygous recessive	White flowers

When Mendel crossed true breeds in the P generation (as shown in the Punnett square below), he noted that all the F_1 offspring had the dominant phenotype. When he crossed F1 offspring (all heterozygotes), the F2 generation consistently showed a 3:1 phenotypic ratio of purple to white flowers. These results demonstrate the first law of Mendelian genetics: the law of dominance.

Parent (P) generation: True breed cross: PP x pp

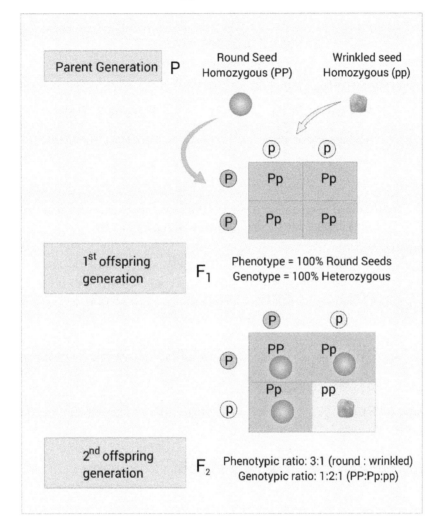

A useful application of this rule is the practice of *backcrossing*. This is when a dominant expressing organism of an unknown genotype is crossed with a recessive expressing organism. Offspring phenotype can be used to determine if the parent is homozygous or heterozygous.

Dihybrid crosses involve two traits and illustrate Mendel's law of segregation: alleles separate in meiosis.

Dihybrid crosses example: Cross PpRr x PpRr

- Purple flowers = P (dominant)
- White flowers = p (recessive)
- Round seeds = R (dominant)
- Wrinkled seeds = r (recessive)

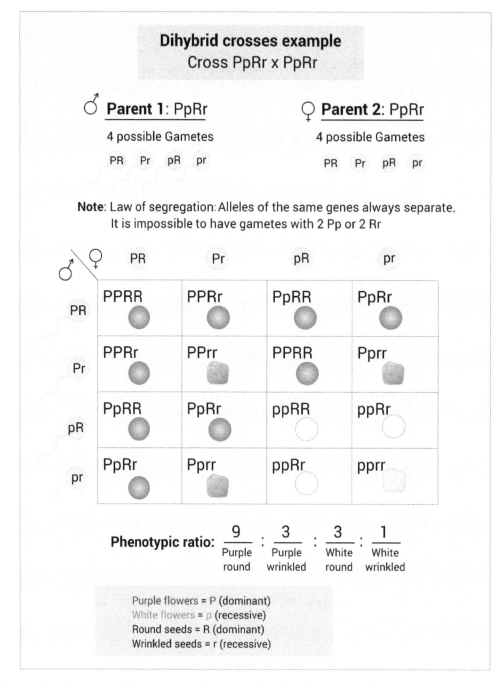

Probability can be determined by using the *law of multiplication* when all factors are present. The *law of addition* can be used to determine probability when one factor OR others may be present. Another dominant phenotype is yellow seed (Y) over green seed (y), and tall stems (T) are dominant over dwarf stems (t).

Question: What is the probability of having offspring with the genotype PPrrYytt if the parent cross is PprrYyTt x PpRryyTt?

Solution:

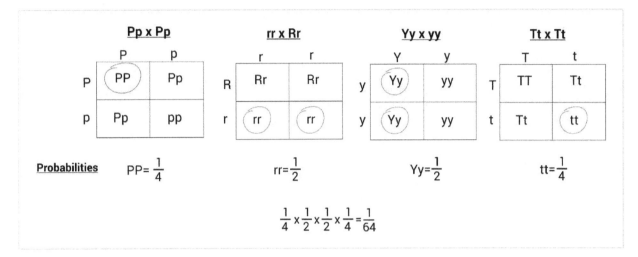

Law of Addition (Mutually Exclusive Results)

Question: What is the probability of having offspring recessive for three different traits (could contain several different combinations)?

Solution:

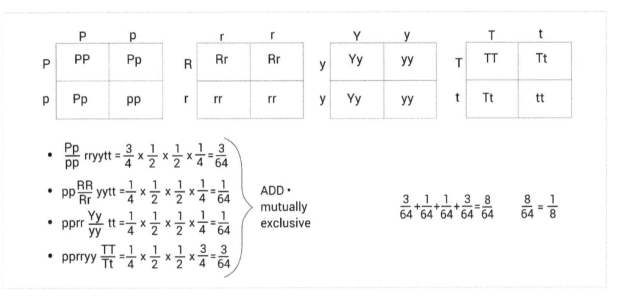

Genetic Crosses

Genetic crosses are the possible combinations of alleles, and can be represented using Punnett squares. A monohybrid cross refers to a cross involving only one trait. Typically, the ratio is 3:1 (DD, Dd, Dd, dd), which is the ratio of dominant gene manifestation to recessive gene manifestation. This ratio occurs when both parents have a pair of dominant and recessive genes. If one parent has a pair of dominant genes (DD) and the other has a pair of recessive (dd) genes, the recessive trait cannot be expressed in the next generation because the resulting crosses all have the Dd genotype. A dihybrid cross refers to one involving more than one trait, which means more combinations are possible. The ratio of genotypes for a dihybrid cross is 9:3:3:1 when the traits are not linked. The ratio for incomplete dominance is 1:2:1, which corresponds to dominant, mixed, and recessive phenotypes.

Co-Dominance and Multiple Alleles

The simple dominant/recessive model for genetics does not work for many genes.

For example, blood type is a trait that has multiple alleles: I^A, I^B, and i. I^A and I^B are "*co-dominant*" so neither is "stronger" than the other, and i is recessive to both. In the event that both co-dominant alleles are present in a genotype, both phenotypes will be present.

Genotype	Phenotype	Blood Donation Facts
IAIA, IAi	A blood (A antigens and B antibodies)	People with A blood can't receive blood from AB or B due to antibody recognition and attack of B antigen.
IAIB	AB blood (A and B antigens but no antibodies)	Universal receiver because it contains no antibodies against A or B antigens.
IBIB, IBi	B blood (B antigens and A antibodies)	People with B blood can't receive blood from AB or A due to antibody recognition and attack of A antigen.
ii	O blood (A and B antibodies)	Can only receive from other O blood (universal donor)

Blood type demonstrates the concept of co-dominance as well as multiple alleles. Below are some blood type crosses and probabilities.

$$\underline{I^A I^A} \times \underline{I^A i} \qquad\qquad \underline{I^A i} \times \underline{I^A i}$$

	I^A	I^A
I^A	$I^A I^A$ A blood	$I^A I^A$ A blood
i	$I^A i$ A blood	$I^A i$ A blood

	I^A	i
I^A	$I^A I^A$ A blood	$I^A i$ A blood
i	$I^A i$ A blood	i i 0 blood

Incomplete dominance occurs when the phenotype is a blending of the two alleles instead of one being dominant over the other. For example, if black and white feathers are co-dominant in birds, heterozygous offspring will have black and white speckles. However, if black and white feathers have an incomplete dominant pattern, heterozygotes will appear grey.

Sex-Linked Genetics

Normal Human Karyotype

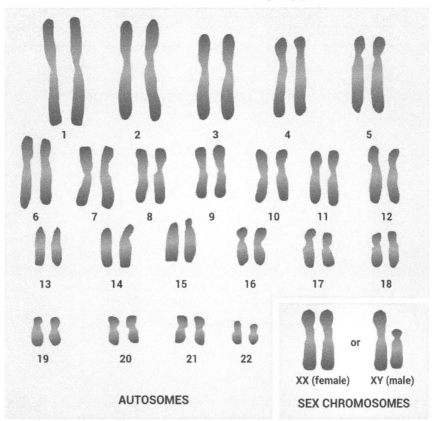

Chromosome 23 is not always homologous because it determines gender, and an individual is either XX (female) or XY (male).

The X chromosome is much larger than the Y chromosome and carries more genes, including the color-blind recessive allele. The possible genotypes, phenotypes, and inheritance patterns are shown below for the color-blind trait:

Genotype	Gender	Phenotype
XC	Female	Normal vision
XC	Female	Normal vision *She carries the allele and can pass it on to her children
Xc	Female	Color blind
XCY	Male	Normal vision
XcY	Male	Color blind

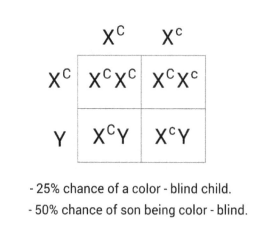

- 25% chance of a color - blind child.
- 50% chance of son being color - blind.

Pleiotropy

Pleiotropy occurs when one gene has more than one phenotype. For example, cystic fibrosis is a debilitating disease that results in mucus overproduction and symptoms that include difficulty breathing and digestive problems.

Polygenic Genes

A phenotype dependent on more than one gene is called *polygenic inheritance*. It is evident when there is a distribution of phenotypes over a wide range, such as skin color. This is in contrast to *monogenic inheritance* when a phenotype is dependent on only one gene.

Epistasis

Epistasis occurs when one gene alters the phenotype of a different gene. For example, in certain mice fur color is determined by the presence of an *Agouti* allele. It codes for brown fur, which is dominant over black fur. Color is also determined by a second allele that codes for pigmentation. The dominant allele C means that fur pigment will be made while the recessive allele c codes for *albinism* (absence of pigment). The gene for pigment disposition can silence the gene for fur color. Possible genotypes/phenotypes are:

- AACC, AaCc = Brown fur
- aaCC, aaCc = Black fur
- AAcc, Aacc, aacc = White fur

Consider the following possible combinations resulting from epistasis of two brown mice, each with the genotype AaCc:

	AC	aC	Ac	Ac
AC	AACC brown	AaCC brown	AACc brown	AaCc brown
aC	AaCC brown	aaCC black	AaCc brown	aaCc black
Ac	AACc brown	AaCc brown	AAcc white	Aacc white
ac	AaCc brown	aaCc black	Aacc white	Aacc white

The resultant phenotypic ratios are as follows: white (4/16), black (3/16), brown (9/16).

Pedigrees

Pedigrees show family ancestry and can be used to track genetic diseases through phenotypes. Circles represent females and squares represent males. Shaded shapes represent affected individuals. The genotypes of individuals can be deduced from given phenotypes. For example, below is a pedigree tracing color blindness. Individual I-1 must be X^cY because he is affected. Individual II-6 must be X^cX^c because, in order for her son to be affected, she had to pass along the color-blind allele she received from her father.

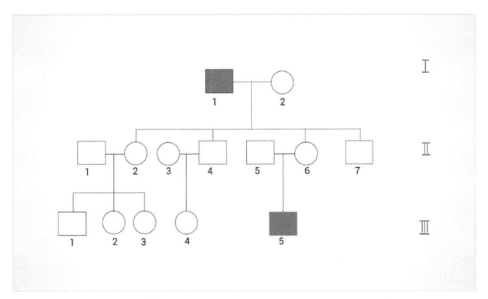

There are two types of *autosomal* or non-sex chromosomes (chromosomes 1-22): *dominant* and *recessive*.

The following pedigree tracing Huntington's disease is an example of an autosomal dominant disease. Every individual that has the dominant allele is affected.

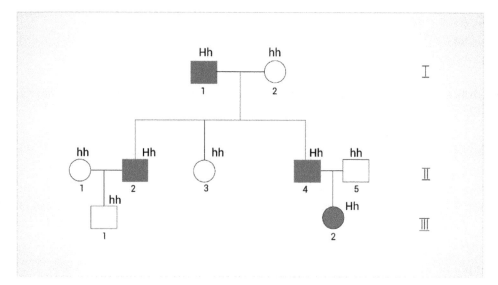

This pedigree tracing albinism is an example of an autosomal recessive disease.

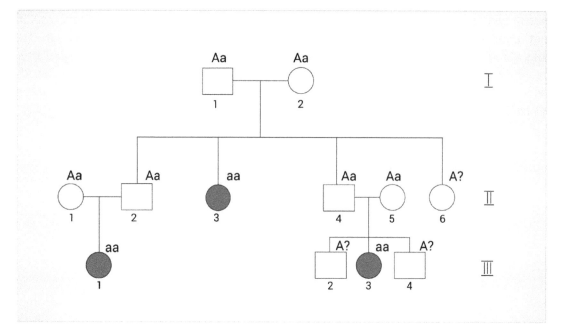

Analysis of this pedigree shows that:

- II-3, III-1, and III-3 have to be aa because they are affected.
- Any couple with children who are affected must both be heterozygotes.
- This particular pedigree can't be a sex-linked, recessive pedigree. In order to be one, an affected female must have an affected father so she can inherit his recessive allele.

Karyotypes

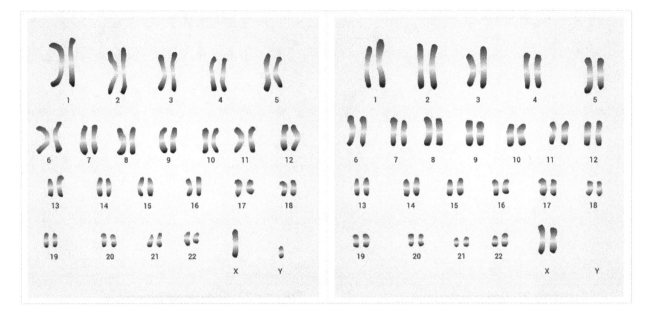

Karyotypes show a picture of an individual's 23 chromosomes and illustrate the diploid nature of our genome. Although females have two X chromosomes, only one X will be active in each cell and the other will form an inactive *Barr Body*. The inactive X chromosome is random. Some cells will have one inactive X chromosome while others have the second X chromosome inactive, resulting in an individual consisting of a mosaic of cells.

Karyotypes not only show gender, they can also illustrate occasional mistakes that occur in meiosis called *nondisjunction*. Nondisjunction results in improper separation of tetrads and chromatids. This results in one cell having an extra copy of a chromosome (*trisomy*) and another cell missing a copy of the homologous chromosome (*monosomy*). Fertilization with gametes affected by nondisjunction is often fatal, but some are viable.

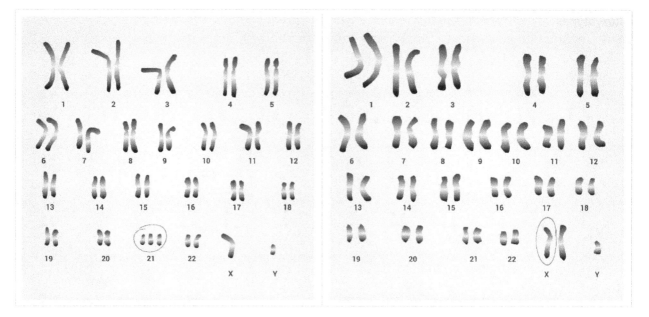

Trisomy 21: Down Syndrome   **Trisomy 23: Klinefelter Syndrome**

Different diseases and their inheritance patterns are listed below:

Disease	Phenotype	Cause/Pattern
Color Blind	Red/green vision deficiency	Sex-linked recessive
Hemophilia	Blood clotting disorder	Sex-linked recessive
Muscular Dystrophy	Weak muscles and poor muscle coordination	Sex-linked recessive
Huntington's	Nervous system degeneration that has a late onset (middle age)	Autosomal dominant
Achondroplasia	Dwarfism	Autosomal dominant
Cystic Fibrosis	Excessive mucus production	Autosomal recessive
PKU	Unable to digest phenylalanine	Autosomal recessive
Tay-Sachs Disease	Intellectually disabled due to inability to metabolize lipids, death in infancy	Autosomal recessive
Sickle Cell Anemia	Red blood cell shaped like a sickle instead of a circle	Autosomal co-dominant
	Disorders caused by nondisjunction and chromosomal deletions:	
Down Syndrome	Intellectually disabled	Trisomy 21
Klinefelter Syndrome	Sterile males	Trisomy 23 (XXY)
Turner Syndrome	Sterile females	Monosomy 21 (x)

Decoding DNA Instructions

Two steps of protein synthesis decode DNA instructions: *Transcription* and *Translation*.

Transcription
Prokaryotic transcription consists of the following three stages:

1. Initiation: RNA polymerase binds to the start site on the RNA template strand, which is upstream in the promotor region of the gene.

2. Elongation: Synthesis of the new complementary RNA strand begins by the action of RNA polymerase as it works its way down the gene, growing the new strand.

3. Termination: The new RNA chain is completed and the RNA polymerase is released at the stop point, which is downstream on the gene in the terminator region.

Eukaryotic transcription is more complicated:

1. Eukaryotic Initiation: Eukaryotes contain advanced promoters, which are DNA sequences upstream of genes that recruit transcription factors, proteins that facilitate or block RNA polymerase binding. The TATA box is where RNA polymerase II binds.

2. Eukaryotic Elongation: This is similar to prokaryotic elongation in that it generates a 5' to 3' complementary RNA transcript.

3. Eukaryotic Termination: A polyadenylation signal (DNA sequence AAUAAA) causes RNA polymerase to release several bases downstream. The RNA transcript is then processed. Upstream of the promoter, past a 5' untranslated region, a "cap" is added (like a telomere, but only a few repeated Guanines). Downstream of the polyadenylation signal, a poly(A) tail is added. In addition to changes at the end of each transcript, areas within the transcript are further processed through splicing. Spliceosomes (large complexes of RNA and proteins) remove intermittent noncoding regions called introns, and the exons (coding sequences) are joined together. The introns are spliced out so that only the exons are part of the final transcript. Alternative splicing (when one transcript is spliced in different ways and creates different proteins) makes this process even more complicated.

Translation
Translation is the process of generating protein from RNA. Translation in prokaryotes is far simpler than in eukaryotes. They only have one circular chromosome (so far fewer genes), and they don't have DNA within a nucleus. They also don't process transcripts. Translation can even occur simultaneously with transcription on the same piece of RNA. In fact, many different ribosomes can be working on the same transcript at the same time, thus creating a structure called a *polyribosome*.

The ribosomes of bacteria and eukaryotes are similar. A ribosome has two subunits, both made of ribosomal RNA (*rRNA*, which is made by RNA polymerase I in eu karyotes). In eukaryotes, the small subunit and the large subunit assemble in conjunction with the AUG start codon in the mRNA message (DNA is read in three letter "words" called codons).

Before translation can start, a transfer RNA molecule (*tRNA*, which is made by RNA polymerase III) must join the ribosomal complex. tRNAs contain anticodons that are complementary to specific codons on the

RNA transcript. On one side, their highly specific anticodon binds to corresponding codons in the mRNA. On the other side, tRNA molecules carry specific amino acids, the monomer of proteins. Once a tRNA molecule has released its amino acid, an enzyme amino-acyl-tRNA synthetase will join a free-floating tRNA and its corresponding amino acid. As a result, the tRNA can continue to deliver fresh amino acids to the ribosomes.

The point of translation is that, every time a tRNA molecule drops off an amino acid, it contributes to an emerging protein. Only when a stop codon is reached will the ribosome disassemble, thus releasing the assembled protein.

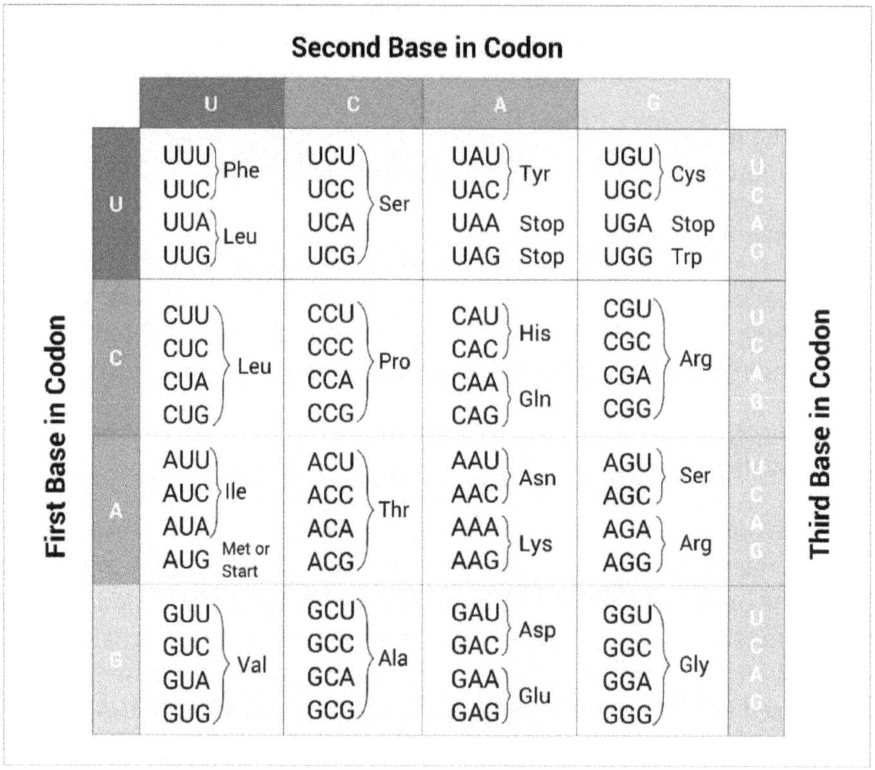

	Second Base in Codon				
	U	C	A	G	
U	UUU UUC Phe / UUA UUG Leu	UCU UCC UCA UCG Ser	UAU UAC Tyr / UAA Stop / UAG Stop	UGU UGC Cys / UGA Stop / UGG Trp	U C A G
C	CUU CUC CUA CUG Leu	CCU CCC CCA CCG Pro	CAU CAC His / CAA CAG Gln	CGU CGC CGA CGG Arg	U C A G
A	AUU AUC Ile / AUA / AUG Met or Start	ACU ACC ACA ACG Thr	AAU AAC Asn / AAA AAG Lys	AGU AGC Ser / AGA AGG Arg	U C A G
G	GUU GUC GUA GUG Val	GCU GCC GCA GCG Ala	GAU GAC Asp / GAA GAG Glu	GGU GGC GGA GGG Gly	U C A G

(First Base in Codon — left axis; Third Base in Codon — right axis)

Note in the chart above that each codon codes for a specific amino acid, even specifically coding for stop codons. There are many codon combinations, but only 20 amino acids. The redundancy in codon/amino acid pairs is due to the third base—or "*wobble*" position—of tRNA. The first two bases bind so strongly that sometimes the third base does not play much of an active role in connecting the codon and anticodon.

The ribosome has three tRNA landing spots. The first tRNA binds to the start codon in the "P" site, but every tRNA that follows lands at the "A" site. The growing amino acid chain is added to the amino acid in the A site (connecting via a peptide bond). The "E" site is where the naked tRNA exits after it has removed its amino acid chain.

Once a stop codon is reached, the amino acid chain leaves through an exit tunnel. At this point, it is an immature protein that is not properly folded and might be sent to the Golgi for modification.

Regulation of Expression

Transcriptional regulation in prokaryotes occurs via operons. An *operon* is a DNA sequence comprised of related genes which are clustered behind a single promoter that regulates them. Within the promoter is an operator, which is like an on/off switch.

Levels of cAMP Regulate Bacterial Gene Expression

The *lac operon* is inducible and controls the digestion of lactose. If lactose is absent, there is no need to make proteins to digest it. A repressor protein will then bind to the operator and block RNA polymerase from transcribing the genes. However, if lactose is present, it will bind to the repressor protein and change its shape, removing the repressor from the operator. The operator will no longer have an interfering protein to stop RNA polymerase, and the proteins that digest lactose will be expressed.

To further complicate matters, lactose has an additional regulator called *CAP* that depends on glucose concentration. Since glucose is the preferred energy source, the lac operon will only be induced when lactose is present in the absence of glucose. If glucose is absent, there will be high levels of a molecule called *cyclic AMP (cAMP)* that will activate CAP. The CAP protein will increase attachment of RNA polymerase to the promoter, thus increasing gene production. If glucose is present, cAMP levels are low and the inactive CAP won't activate transcription.

There are also repressible operons, as is the case with the regulation of tryptophan. In this case, the operator is normally on (meaning it has no repressor protein) and the amino acid tryptophan is made. Excess tryptophan forms a repressor protein/tryptophan complex that binds to the operator and blocks RNA polymerase, thus halting transcription.

Transcription Regulation

As discussed earlier, eukaryotic transcription initiation involves transcription factors. Different transcription factors are present in different cell types, and certain sequences of DNA enhance protein binding. This results in an altered DNA spatial arrangement and exposure to RNA polymerase. Some transcription factors facilitate transcription while some block it. The presence or absence of these proteins in certain cells is one way gene production is regulated.

Another way transcription is regulated in eukaryotes is by gene accessibility. DNA is wrapped around histone proteins in complexes called *nucleosomes*. These nucleosomes coil and supercoil, which make portions of the genome inaccessible. Modification of histone tails in nucleosomes open and close regions of DNA, making them either more or less available for the protein binding of transcription factors and RNA polymerase. When the chromatin is in its closed, coiled conformation, called *heterochromatin,* transcription is repressed because the genes are inaccessible. In contrast, in the *euchromatin* formation, the chromatin is open and uncoiled, which allows RNA polymerase to access the genes, which activates transcription.

Essentially, open chromatin (*euchromatin*) has acetylated histone tails with few methyl groups. *Heterochromatin* is the opposite, with histone tails being highly methylated and deacetylated so that DNA is closed and condensed.

MicroRNAs (*miRNAs*) and small interfering RNAs (*siRNAs*) can also regulate gene production. This is done by degrading certain transcripts or blocking their translation, and sometimes even altering chromatin structure.

It should be noted that there is far more DNA functionality than gene expression. DNA also contains promoters and enhancers to regulate gene expression, centromeres, telomeres, transposable elements, and other sequences.

<u>Transposable Elements</u>
These are literally "jumping genes" that relocate within an organism's genome. Prokaryotes and eukaryotes contain *transposons*. This supports the idea that they have a significant effect on biodiversity and arose through common ancestry among all living organisms. DNA sequences prone to transposon trespassing are predictable, repeated bases that can extend from 300 to 30,000 nucleotides.

Within an organism, transposons relocate and might leave the targeted segment behind. A transposon can make a copy of the DNA or simply cut out the sequence. Either way, transposase is a critical enzyme involved in the process.

Retrotransposons are elements that copy the targeted segment into an RNA transcript. The enzyme retrotransposase is then used to copy the intermediate back into DNA before inserting it, thus copying part of the genome and moving it elsewhere.

It's believed that all human ancestors once had brown eyes until a single individual had a mutation which coded for blue eyes. Once that allele entered the gene pool, it was passed to generations of offspring. Today it's fairly common to see a blue-eyed individual. Thus, changes in alleles can change phenotypes.

Variations are Introduced by the Imperfect Nature of DNA Replication and Repair

Despite replication's proofreading, mutations do happen. Mutations can be either point mutations or chromosomal mutations, and both increase genetic variation.

Point mutations involve a change in DNA sequence and are due to an addition, deletion, inversion, or substitution error.

Insertion and deletion mutations cause frame-shift mutations. This means that every following codon will be read incorrectly, massively changing the primary structure of the protein. Due to the redundancy of codon/amino acid pairing resulting from the wobble position, some substitution mutations cause no change in the protein sequence. The result is something called a silent mutation. All point mutations (including substitution and inversion) can cause gene malfunction.

<u>Chromosomal Mutations</u>
Sometimes there are *chromosomal mutations* in DNA replication and mitosis. The following types of mutations can occur:

- *Deletion*: A section of a chromosome is removed.
- *Duplication*: A section of a chromosome is repeated.
- *Inversion*: A chromosome is rearranged within itself.
- *Translocation*: Chromosome pieces mix or combine with other chromosomes.

Not only can mutations lead to changes in an individual's phenotype, but they can also contribute to reproductive isolation and speciation which have effects on a much larger scale.

In metaphase I in meiosis, homologous chromosomes align in the center of the cell. The law of independent assortment states that it is random whether a mother's or father's chromosomes are on the left or the right. Since humans have 23 chromosomes, there are 2^{23} (over 8 million) possible variations. Independent assortment, random fertilization, and recombination contribute greatly to genetic diversity.

An Exception to Independent Assortment: Gene Linkage

A man named Thomas Hunt Morgan did extensive genetic research using drosophila (flies), which are an effective test subject because they only have four chromosomes. Using the phenotypes of body color and eye color (grey body and red eyes are dominant), below is a dihybrid cross illustrating the expected phenotypic ratios.

ggrr x GgRr

	gr	gr	gr	gr
GR	GgRr	GgRr	GgRr	GgRr
Gr	Ggrr	Ggrr	Ggrr	Ggrr
gR	ggRr	ggRr	ggRr	ggRr
gr	ggrr	ggrr	ggrr	ggrr

Phenotypic ratio: 4 : 4 : 4 : 4

$$\frac{4}{\substack{\text{gray body}\\\text{red eyes}}} : \frac{4}{\substack{\text{gray body}\\\text{purple eyes}}} : \frac{4}{\substack{\text{black body}\\\text{red eyes}}} : \frac{4}{\substack{\text{black body}\\\text{purple eyes}}}$$

However, what Morgan actually observed was not consistent with the expected results. He observed that the cross ggrr x GgRr had a highly skewed phenotypic ratio. Today we understand that this occurs because the two genes are located on the same chromosome and, as such, are more likely to be passed together. Genes close to each other are "linked" and infrequently separate in recombination.

There is a mathematical formula used to find recombination frequency and develop a linkage map illustrating the genes' relative locations on a chromosome. Here is that formula:

$$\frac{Number\ of\ recombinants}{Total\ offspring} \times 100 = Recombination\ frequency$$

113

Given the calculation results, a linkage map can be constructed to show the relative location since the closer the genes the lower the recombination frequency.

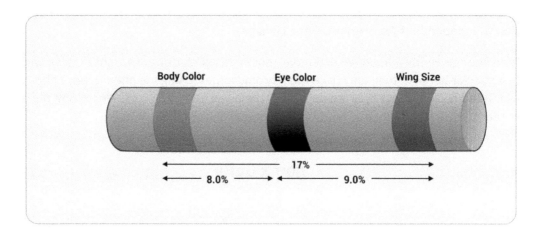

Bacterial Genetic Variation

Occasionally bacteria do have DNA mutations, but they don't harbor chromosomal mutations because they only have one large, central chromosome. They don't perform recombination, instead reproducing asexually via binary fission. However, they can still have genetic variation through the following processes:

1. Transformation

2. Transduction

3. Conjugation

4. Mutation

5. Transposable element relocation

Transformation	Transduction	Conjugation
Exogenous snippets of DNA enter bacterial cells.	Bacteriophages (viruses that infect bacteria) introduce foreign DNA into a bacterial cell.	One bacterial cell extends a pilus into another and releases DNA.

Transformation	Transduction	Conjugation
Exogenous snippets of DNA enter bacterial cells.	Bacteriophages (viruses that infect bacteria) introduce foreign DNA into a bacterial cell	One bacterial cell extends a pili into another and releases DNA.

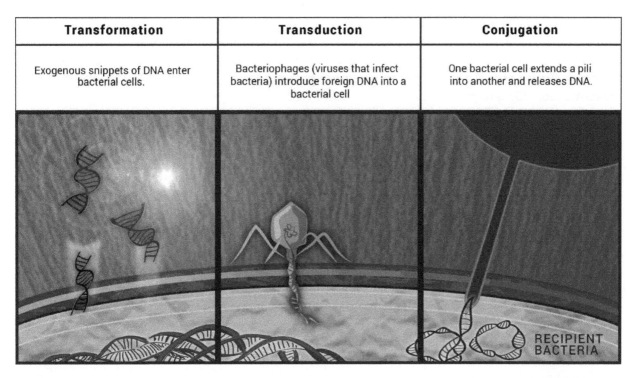

In conjugation, one bacterial cell is the giver (F$^+$) while the other is the receiver (F$^-$). After the pilus extends and attaches to the recipient, it inserts in F factor, a plasmid containing the new allele. Sometimes R plasmids, which contain and spread antibiotic-resistant genes, are transferred.

Viral Replication

Viruses were discussed above as inducers of genetic variation in bacteria via transduction. Viruses (bacteriophages or simply phages) infect bacteria via the lytic or lysogenic cycle. Both processes begin with the bacteriophage injecting its viral genome into a host cell.

Lytic Cycle: This cycle occurs when the viral DNA immediately takes over the host cell, tricking it into copying and translating it. The host cell has no way of telling the difference between the viral and self DNA. By the end of the lytic cycle, hundreds of viruses have been produced and released from the bacterial cell when it "lyses" or bursts.

Lysogenic Cycle: Once the viral DNA enters, it doesn't immediately direct cellular activities. Instead, it embeds itself into the host cell DNA and is called a prophage. This prophage "hides" in the host cell DNA and, as the bacteria replicates via binary fission, it passes its virulent genes on to future generations of cells. Eventually, a signal will activate the viral DNA. When this occurs, the prophage will direct the cell to initiate the lytic cycle, making its host cell a virus-making factory.

Bacteriophages consist of DNA surrounded by a protein coat called a *capsid*. However, they aren't alive because they don't have ribosomes and, thus, can't grow or reproduce on their own.

Like bacteriophages, animal viruses contain nucleic acid surrounded by a capsid. They often contain a viral envelope or an outer membrane. Animal viruses can contain DNA as their genetic material, while some contain RNA instead. Those containing RNA may also carry an enzyme called *reverse transcriptase*, which allows their transcript to be copied into DNA in the host cell's cytoplasm.

Like all viruses, HIV has specificity as to which cells it can infect. This is due to cell recognition. HIV contains certain marker proteins or *antigens* which are only recognized by a subset of cells, primarily helper T cells. These *helper T cells* are activators of the entire specific immune response that stimulates antibody production. Should the helper T cell count drop, an individual develops Acquired Immune Deficiency Syndrome (AIDS), a devastating disease that destroys the immune system. If enough T cells are destroyed, AIDS is fatal.

The table below briefly describes the steps in which the HIV retrovirus infects a helper T cell:

Step	Explanation
1	Antigens on the HIV virus are recognized by helper T cell receptors and HIV binds to the surface of the host cell.
2	Host ingests viral genome and proteins, including reverse transcriptase.
3	Reverse transcriptase copies Viral RNA and makes Viral DNA.
4	Viral DNA integrates into the cellular DNA, forming a provirus. Like a prophage, the Viral DNA can be inactive for a long period of time (called latency).
5	Viral RNA is produced and host cell ribosomes translate its instructions.
6	Viruses assemble at the cell surface and new immature copies of HIV form.
7	The new HIV matures and exocytosis occurs. The vrus particles are released from the cell.

Viroids and Prions

Even smaller than viruses are simple structures call viroids and prions. *Viroids* are mini-viruses (nucleotides without protein) that specifically infect plants. *Prions* are misfolded proteins that infect animals. They affect brain function and cause diseases like mad cow.

Communication between Cells

All cells have some form of cellular communication, even prokaryotes. Bacteria secrete small molecules that signal area density. This is known as *quorum sensing*.

Even single-celled yeast have signal transduction pathways similar to those of animals, and the conservation suggests a shared ancestor.

Multicellular organisms have the following junctions involved in cellular communication:

Gap Junctions: Pores surrounded by membrane proteins that exist between animal cells, thus allowing small molecules to pass through.

Plasmodesmata: Membrane-lined channels between the cell walls of adjacent plants

Animals can elicit localized responses from signals produced by a single cell via diffusion. This is called *paracrine signaling*, and it's an important process in both the immune response and development. Also localized is the nervous system signaling, since one activated neuron stimulates the next by neurotransmitter release.

Distance signaling delivers substances like hormones to cells via the circulatory system in animals and via plasmodesmata and diffusion in plants. Hormones, ions, and other signals (together called *ligands*) bind to receptor proteins on their specific target cells. Through a series of steps, they elicit a cellular response.

There are three main types of receptors: *G protein-coupled receptors, enzyme-linked receptors*, and *ion channel receptors*. All three initiate their response upon ligand binding.

- Activated G proteins add a phosphate to an inactive G protein (involving GTP instead of ATP). The activated G protein binds to and activates a nearby enzyme that initiates a signal transduction cascade.

- Activated enzyme-linked receptors are transmembrane receptors that induce phosphorylation of cytoplasmic kinases, initiating a signal transduction cascade. Examples include receptor tyrosine kinases and receptor serine/threonine kinases.

- Activated ion-gated channels open for ion transport. The displaced ions initiate a signal transduction cascade.

Transduction
Regardless of the type of receptor, once the ligand binds it starts a series of events that affect the cell and are collectively referred to as the *phosphorylation cascade*.

The signal transduction pathway is a tightly regulated relay race. It involves activating a molecule that changes its shape and activates its neighbor. After this activating activity, it resumes its original inactive shape.

Kinases are proteins that phosphorylate and phosphatases de-phosphorylate. Their interrelationship is critical for the quick activation and inactivation of involved proteins.

Sometimes the players involved in signal transduction are not proteins. They are called *second messengers*. The first messenger is the ligand. The second messenger, often cyclic AMP or Calcium ions, is produced quickly and in large amounts.

Some nonpolar ligands, like steroids, are permeable to the membrane so they don't bind to receptors. Instead, they diffuse to the nucleus where they act as transcription factors to regulate gene expression.

Cellular Response
The response can occur at the transcriptional level by increasing or decreasing the production of transcription factors and genes. It can also occur by stopping, starting, or fine-tuning protein production via protein synthesis.

The response can occur in the cytoplasm by affecting the target of the protein, such as membrane channel behavior.

The nature of the relay race is such that each step activates more products than the step before. If one protein activates two others, the next two proteins will have the same window of time and each will activate two (for a total of four). This cascade and amplification effect leads to a lightning-quick response.

Termination
Ligand binding is reversible. If there is a high level of external ligand, the system will be activated. If not, the system will be deactivated.

Humans Can Act on Information and Communicate it to Others

Thanks to their nervous system and complex brain, humans can process information and communicate.

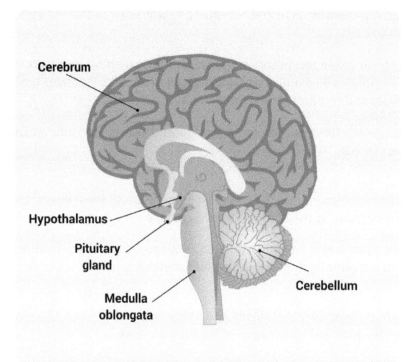

The human brain can be divided into different areas. An overview of their functionality is listed below:

Cerebrum: The cerebrum is responsible for voluntary movement, learning, emotion, and advanced cognition. The right and left hemisphere are connected by the corpus callosum.

Cerebellum: The cerebellum is responsible for coordination and balance. The base of cerebellum contains amygdala which is important for emotions.

Diencephalon: This contains the thalamus and hypothalamus. The hypothalamus controls temperature, acts as a biological clock, and controls the pituitary gland (hormone regulator).

Brainstem: The brainstem contains the medulla oblongata and pons. The medulla oblongata controls involuntary homeostatic functions (breathing and heart rate). The pons delivers sensory information to the midbrain and forebrain.

The Nervous System

The nervous system can be broken down into the following categories:

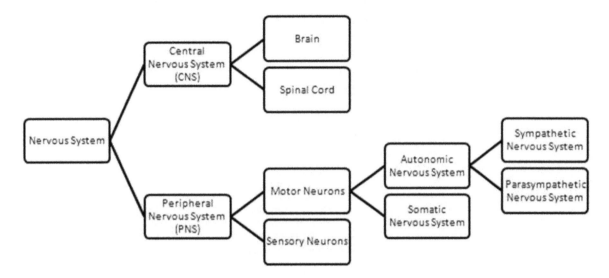

Neurons: Building Blocks of the Nervous System

There are several types of neurons:

Sensory Neurons: These neurons sense external stimuli (sights, sounds, smells, tastes, and touches) and internal stimuli (temperature, blood sugar levels, blood pressure).

Interneurons: Neurons that are involved in the brain circuitry which relays signals between sensory or motor neurons and the Central Nervous System.

Motor Neurons: These neurons stimulate reactions in the muscle cells.

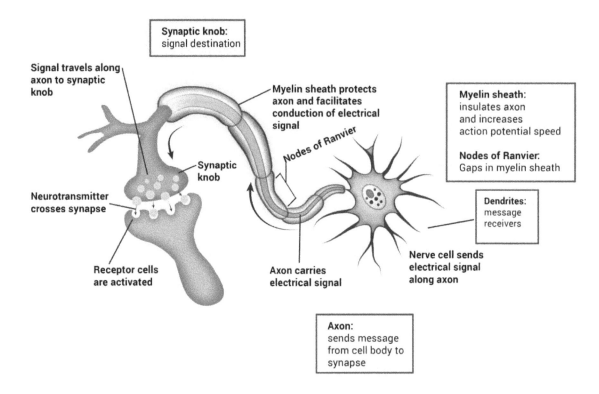

Ion-gated channels are critical for action potentials to occur. A cell's resting membrane potential is actually negative, but if a sodium ion channel is opened, Na$^+$ ions will rush in and depolarize the cell. Alternatively, if a potassium ion channel is opened, K$^+$ ions will rush out and hyperpolarize the cell.

An action potential involves the following steps:

1. Axon must be stimulated past its threshold.

2. Na+ gates open and Na+ ions rush in and depolarize the membrane.

3. K+ gates then open and K+ ions rush out.

4. This polarity change is called an action potential. It is very brief because the sodium potassium pump quickly restores the resting potential. The sodium potassium pump is an ATP-driven transmembrane protein that delivers three Na+ ions out of the cell and two K+ ions into the cell.

5. Neurons can't immediately fire again because they have a refractory period. This is due to the brief period of time it takes to return to resting potential.

6. The signal passes from axons of the presynaptic cell to dendrites of the postsynaptic cell via neurotransmitters at the synaptic knob.

The synaptic transmission is different from the electrical, voltage-dependent method above.

1. Presynaptic membrane depolarized as Ca^+ ions rush in from calcium-gated channels.

2. Neurotransmitters within vesicles are delivered by exocytosis into the synapse.

3. Neurotransmitters bond with receptors on the postsynaptic membrane.

4. The postsynaptic cell is either activated (depolarized) or inhibited (hyperpolarized) as a response.

5. Enzymes recycle neurotransmitters floating in the synaptic knob, and they are returned to the presynaptic neuron.

Common Neurotransmitters:

- Acetylcholine
- Serotonin
- Epinephrine
- Norepinephrine
- Dopamine
- Glutamate
- GABA

Practice Questions

1. A color-blind male and a carrier female have three children. What is the probability that they are all color blind?
    a. 1/4
    b. 1/8
    c. 1/16
    d. 1/32

Use the image below to answer questions 2-3:

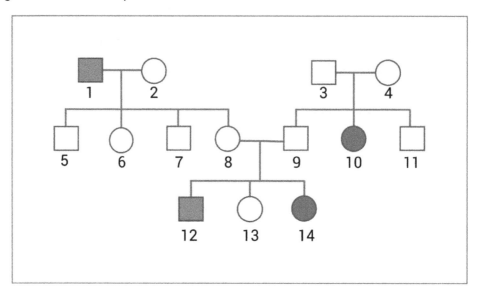

2. What kind of pedigree is shown?
    a. Autosomal dominant
    b. Autosomal recessive
    c. Sex-linked dominant
    d. Sex-linked recessive

3. What is the genotype of individual 9?
    a. AA
    b. Aa
    c. aa
    d. $X^A Y$

4. This karyotype indicates what about the individual?

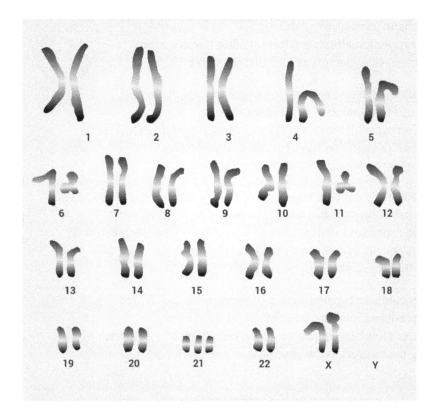

    I. They are female
    II. They are male
    III. They have Down Syndrome
    IV. They have Turner Syndrome

    a. I and III
    b. I and IV
    c. II and III
    d. II and IV

5. What is the probability of AaBbCcDd x AAbbccDD having a child with genotype AAbbccDD?
    a. 1/2
    b. 1/4
    c. 1/8
    d. 1/16

6. If a person with AB blood and a person with O blood have children, what is the probability that their children will have the same phenotype as either parent?
    a. 0%
    b. 25%
    c. 50%
    d. 75%

7. Which of the following occurs in epistasis?
    a. There are two alleles: dominant and recessive
    b. Two alleles share dominance
    c. One allele's expression affects a different allele's expression
    d. Heterozygotes have a phenotype of blended alleles

8. The law of independent assortment states which of the following?
    a. Tetrads line up in metaphase I in a random fashion
    b. Nondisjunction occurs if there are errors in anaphase I or anaphase II
    c. Sperm will fertilize eggs that have certain characteristics
    d. Dominant alleles mask recessive allele phenotypes

9. Which of the following is *not* true regarding cell cycle checkpoints in mitosis?
    a. A cyclin/cdk pair is responsible for assembling mitotic machinery.
    b. Cyclin protein increases in interphase and is broken down during mitosis.
    c. Cdk protein is equally expressed throughout the cell cycle.
    d. G_0 is a state that stimulates progression into S phase.

10. Which statements regarding meiosis are correct?
    I. Meiosis produces four diploid cells.
    II. Meiosis contains two cellular divisions separated by interphase II.
    III. Crossing over occurs in the prophase of meiosis I.

    a. I and II
    b. I and III
    c. II and III
    d. III only

11. Given the following recombination frequencies, which linkage map would be correct?
    D-A = 9%
    D-B = 3%
    B-C = 5%
    C-A = 1%
    a. ACBD or DBCA
    b. DCAB or BACD
    c. BADC or CDAB
    d. ABCD or DCBA

12. Which of the following is true regarding the lac operon in prokaryotes?
    a. It is repressible
    b. It is blocked by a repressor that binds to the lacZ site in the absence of lactose
    c. It regulates one gene important for lactose digestion
    d. It is also regulated by the presence of glucose and cAMP

13. Which of the following sequences are significant during transcription of eukaryotic mRNA?
  - I. TATA
  - II. AAAUAA
  - III. AUG

  a. I only
  b. II only
  c. III only
  d. I and II

14. In protein synthesis, which of the following is *not* true about the molecule tRNA?
  a. It obtains its amino acid by aminoacyl-tRNA synthetase
  b. It has an anticodon that is complementary to an mRNA codon
  c. The wobble position is the third base in the anticodon
  d. It contains a large subunit and a small subunit

15. DNA must be created in a 5′ to 3′ direction. This causes which of the following?
  a. Shortening of telomeres
  b. Hydrogen bonds between nitrogen bases
  c. Primase binding
  d. Okazaki fragments to form on the leading strand

16. Which phenomenon is *not* involved in prokaryotic genetic diversity?
  a. Mutation
  b. Transformation
  c. Conjugation
  d. Crossing over

17. Which statement is not true regarding the lytic and lysogenic cycles?
  a. Both processes begin with phage attachment.
  b. Only the lysogenic cycle applies to eukaryotic cells.
  c. Lysogenic is dormant until an activating factor stimulates viral production.
  d. Both cycles utilize host cell ribosomes to make more viruses.

18. Which event occurs first in receptor-mediated signal transduction involving receptors?
  a. Phosphorylation/dephosphorylation cascade
  b. Transcription factor activation by activated protein
  c. Ligand binding to a tyrosine kinase transmembrane protein
  d. Ion-gated channels open

19. Which of the following could be a second messenger in a signal transduction pathway?
  a. G-coupled protein receptor
  b. Transcription factor
  c. Cyclic AMP
  d. Ligand

20. Which process occurs immediately after an action potential?
    a. Sodium ions rush into the cell
    b. Calcium ions rush out of the cell
    c. Hyperpolarization
    d. Potassium ions rush into the cell

Answer Explanations

1. B: 1/8. Color blindness is a recessive, sex-linked trait. The Punnett square below shows the cross between a carrier female and a color-blind male. The two offspring in bold are color blind. The probability of having one child that is color blind is ½. The probability of having three color-blind children is ½ x ½ x ½ (law of multiplication) or 1/8.

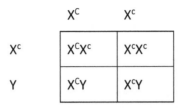

2. B: Autosomal recessive. For dominant pedigrees, it would be impossible for two recessive parents to have a child that expresses the dominant trait, as is seen in individual 10, making Choices *A* and *C* wrong. This cannot be a sex-linked recessive pedigree because of individual 14: a girl cannot be color blind unless her father is color blind, making Choice *D* incorrect (see Punnett square from #1). The correct answer is *B*, autosomal recessive.

3. B: Aa. This is an autosomal recessive pedigree, so Choice *D* is incorrect. Individual 9 has a child who has the trait, so he must have a recessive allele. He must also have the dominant allele since he does not have the trait. Choice *B* is the heterozygous genotype that has both the dominant and recessive allele, so the correct answer is *B*.

4. A: I and III. This is a female because her 23rd chromosome pair is composed of two X chromosomes and no Y. The karyotype also shows trisomy 21, which is Down syndrome. Turner syndrome is monosomy 23 (women with only one sex chromosome), making IV incorrect.

5. D: 1/16. The probability of each specified genotype can be determined by individual Punnett squares. Each probability should then be then multiplied (law of multiplication) to find the value, which in this case is ½ x ½ x ½ x ½ = 1/16.

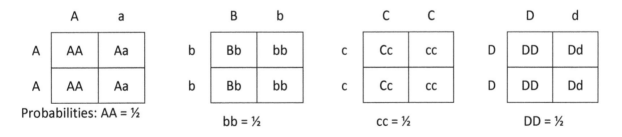

6. A: 0%. There is no chance that an offspring will be O blood or AB blood (see Punnett square).

	IA	IB
i	IAi	IBi
i	IAi	IBi

7. C: One allele's expression affects a different allele's expression. Co-dominance and incomplete dominance are described in Choices *B* and *D*, not epistasis, so they are incorrect. Simple Mendelian genetics is described in Choice *A*, which is also incorrect.

8. A: Tetrads line up in metaphase I in a random fashion. Choice *B* is not a heredity issue; nondisjunction is a mistake in chromatid separation that is not inherited. Choice *C* is untrue because fertilization is random. Choice *D* explains alleles, but does not explain the mechanism behind genetic diversity like Choice *A* does. The law of independent assortment pertains to the random lineup of chromosomes in metaphase I.

9. D: G_o is a state that stimulates progression into S phase. G_o is actually a checkpoint that involves cells exiting the cell cycle, such as mature neurons and damaged cells that may undergo apoptosis. Therefore Choice *D* is the untrue statement. Choices *A*, *B*, and *C* refer to the Maturation Promoting Factor (MPF), which is a cyclin/cdk pair that forms in G_2 when rising levels of cyclin bind to and activate an ever-present cyclin dependent kinase.

10. D: III only. Roman numeral I is incorrect because meiosis produces haploid cells. Roman numeral II is incorrect because there is no interphase II (otherwise gametes would be diploid instead of haploid). Choice *D* is the only correct answer because the others contain Roman numerals I and II.

11. A: ACBD or DBCA. This order is the only one that works so the map units have the assigned distances between them.

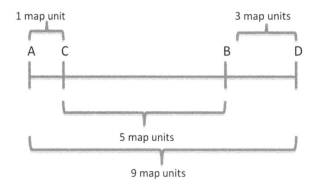

D- A = 9%
D-B = 3%
B-C = 5%
C-A = 1%

12. D: Is also regulated by the presence of glucose and cAMP. Operons in bacteria have a regulatory "switch" (DNA sequence called an operator) that controls transcription of several related genes, so Choice *C* is incorrect. There are three genes for lactose digestion, not one. lac operon is inducible. It is mostly off and only turned on by the presence of lactose, so *A* is incorrect. Similarly, because it is inducible, lactose will bind to the repressor so that RNA polymerase is no longer blocked, so Choice *B* is incorrect. The only correct description of the lac operon is Choice *D*.

13. D: I and II. Roman numeral I, the TATA box, is important for RNA polymerase recruitment. Roman numeral II (AAAUAA) is the poly(A) tail that is a signal to stop transcription. Roman numeral III (AUG) is the start codon, which is an important sequence for translation, not transcription. Therefore, only the first two sequences are important during mRNA transcription.

14. D: It contains a large subunit and a small subunit. rRNA is the type of RNA that has subunits, not tRNA. The other answer choices (*A*, *B*, and *C*) are all true.

15. A: Shortening of telomeres. Choices *B* and *C* both occur, but they have nothing to do with the direction of DNA synthesis. Choice *D* is untrue because Okazaki fragments are due to the directional synthesis of DNA, but they form on the lagging strand.

16. D: Crossing over. All choices cause genetic variation in bacteria except for crossing over, which is strictly a eukaryotic event that occurs with linear, homologous chromosomes.

17. B: Only the lysogenic cycle applies to eukaryotic cells. Choice *C* is true since the lysogenic cycle is dormant until an activating factor stimulates viral production. Choice *A* is true because both start with viral attachment. Choice *D* is true because the lysogenic cycle, once activated, utilizes host cell membranes like the lytic does. Choice *B* is untrue because the lysogenic cycle is specific to prokaryotes and is not a process in eukaryotic viral infections.

18. C: Ligand binding to a tyrosine kinase transmembrane protein. Tyrosine kinase transmembrane proteins are just one example of a receptor protein. There are also G protein-coupled receptors and ion channel receptors. Regardless of the type of receptor, ligand binding is the first step. Choice *D* is incorrect because, prior to ion-gated channels opening, a ligand would need to bind to the receptor to induce the conformational change. Choice *A* occurs after ligand binding and Choice *B* is a response that is much farther downstream (it's an effect of signal transduction).

19. C: Cyclic AMP. A second messenger is a non-protein, so Choices *A* and *B* are incorrect. Choice *D* is incorrect because the ligand is the first messenger.

20. A: Sodium ions rush into the cell. Choice *C* is incorrect because sodium influx causes depolarization, not hyperpolarization. Choice *D* is incorrect because potassium ions actually rush out of the cell. Choice *B* is incorrect because calcium ions are not involved (until later at the synaptic knob).

Interaction of Biological Systems

Monomers Join to Form Polymers

All biological material is made up of atoms bonded together to form molecules. These bonds give the molecules their structure and determine the function of the molecule. Many biologically important molecules are polymers, which are chains of a repeated basic unit, called a monomer.

Nucleic Acids

Two types of five-carbon sugars can form nucleotides: ribose and deoxyribose. Ribose is the sugar found in ribonucleic acid (RNA). Deoxyribose is the sugar found in deoxyribonucleic acid (DNA), and it has one less oxygen atom than ribose. The important nitrogenous bases are adenine, guanine, cytosine, thymine, and uracil. Thymine is found only in DNA and uracil is found only in RNA. The structure of these molecules determines their function. DNA and RNA have different properties because ribose and deoxyribose have slightly different structures. DNA is usually found as a double helix because deoxyribose has more flexibility. Nucleic-acid function is also determined by the sequence and properties of the nitrogenous bases. Nitrogenous bases form hydrogen bonds with specific other nitrogenous bases. Adenine interacts with thymine and uracil, and cytosine interacts with guanine; these interactions are called *base pairs.*

This specific pairing allows DNA to serve as the hereditary material for cells because it can be copied accurately and passed down to daughter cells. DNA is also able to serve as hereditary material because the sequence of the nitrogenous bases acts as a code that can be made into all of the proteins needed by the cell. RNA is transcribed from DNA accurately because the nitrogenous bases from RNA interact with those from DNA. Proteins are translated from this RNA using a special type of RNA, called a transfer RNA. The transfer RNA has three nucleotides that can bind to the RNA being used to make the protein.

Proteins

Proteins are molecules that consist of carbon, hydrogen, oxygen, nitrogen, and other atoms, and they have a wide array of functions. The monomers that make up proteins are amino acids. All amino acids have the same basic structure. They contain an amine group (-NH), a carboxylic acid group (-COOH), and an R group. The R group, also called the functional group, is different in each amino acid.

The functional groups give the different amino acids their unique chemical properties. There are twenty naturally occurring amino acids that can be divided into groups based on their chemical properties. Glycine, alanine, valine, leucine, isoleucine, methionine, phenylalanine, tryptophan, and proline have nonpolar, hydrophobic functional groups. Serine, threonine, cysteine, tyrosine, asparagine, and glutamine have polar functional groups. Arginine, lysine, and histidine have charged functional groups that are basic, and aspartic acid and glutamic acid have charged functional groups that are acidic.

A peptide bond can form between the carboxylic-acid group of one amino acid and the amine group of another amino acid, joining the two amino acids. A long chain of amino acids is called a polypeptide or protein. Because there are so many different amino acids and because they can be arranged in an infinite number of combinations, proteins can have very complex structures. There are four levels of protein structure. Primary structure is the linear sequence of the amino acids; it determines the overall structure of the protein and how the functional groups are positioned in relation to each other, as well as how they interact. Secondary structure is the interaction between different atoms in the backbone chain of the protein. The two main types of secondary structure are the alpha helix and the beta sheet. Alpha helices are formed when the N-H of one amino-acid hydrogen bonds with the C=O of an amino acid four amino acids earlier in the chain.

The functional groups of certain amino acids—including methionine, alanine, uncharged leucine, glutamate, and lysine—make the formation of alpha helices more likely. The functional groups of other amino acids, such as proline and glycine, make the formation of alpha helices less likely. Alpha helices are right-handed and have 3.6 residues per turn. Proteins with alpha helices can span the cell membrane and are often involved in DNA binding. Beta sheets are formed when a protein strand is stretched out, allowing for hydrogen bonding with a neighboring strand.

Similar to alpha helices, certain amino acids have an increased propensity to form beta sheets. Tertiary protein structure forms from the interactions between the different functional groups and gives the protein its overall geometric shape. Interactions that are important for tertiary structure include hydrophobic interactions between nonpolar side groups, hydrogen bonding, salt bridges, and disulfide bonds. Quaternary structure is the interaction that occurs between two different polypeptide chains and involves the same properties as tertiary structure. Only proteins that have more than one chain have quaternary structure.

Lipids

Lipids are a very diverse group of molecules that include fats, oils, waxes, and steroids. Since most lipids are primarily nonpolar, they are hydrophobic; however, some lipids have polar regions, making them amphiphilic, which means they are both hydrophobic and hydrophilic. Because the structure of different lipids is so diverse, they have a wide range of functions, which include energy storage, signaling, structure, protection, and making up the cell membrane. Triglycerides are one type of biologically important lipid. They are made up of one molecule of glycerol bonded to three long fatty-acid chains, which are long hydrocarbon chains with a carboxylic acid group.

Many of the properties of triglycerides are determined by the structure of the fatty-acid chains. Saturated triglycerides are made of hydrocarbon chains that only have single bonds between the carbon atoms. Because this structure can pack in more tightly, these triglycerides are often solids at room temperature. Unsaturated triglycerides have double bonds between one or more pairs of carbon atoms. These double bonds create a kink in the fatty-acid chain, which prevents tight packing of the molecules; thus, triglycerides are often liquid at room temperature.

Nucleic acids are polymers whose biological function is to provide hereditary information for all living creatures. The monomer of nucleic acids is a nucleotide. Nucleotides consist of a five-carbon sugar, a nitrogenous base, and a phosphate group.

Carbohydrates

Carbohydrates are molecules consisting of carbon, oxygen, and hydrogen atoms and serve many biological functions, such as storing energy and providing structural support. The monomer of a carbohydrate is a sugar monosaccharide. Some biologically important sugars include glucose, galactose, and fructose. These monosaccharides are often found as six-membered rings and can join together by a process known as condensation to form a disaccharide, which is a polymer of two sugar units.

Polysaccharides are long chains of monosaccharides. Their main biological functions are to store energy and provide structure. Starch and cellulose are both polymers of glucose made by plants, but their functions are very different because the structure of the polymers are different. Plants make starch to store the energy that is produced from photosynthesis, while cellulose is an important structural component of the cell wall. These two molecules differ in how the molecules of glucose are bonded together. In starch, the bond between the two glucose molecules is a high-energy alpha bond that is easily hydrolyzed, or broken apart. The bonds between the glucose molecules in cellulose are beta bonds. Being stiff, rigid, and hard to hydrolyze, these bonds give the cell wall structural support. Other important carbohydrates include glycogen, which provides energy storage in animal cells, and chitin, which provides structure to arthropod exoskeletons.

Directionality Affects the Structure and Function of a Polymer

The structure of these biological molecules often results in directionality, where one end is different from the other.

In nucleic acids, the five carbons that make up the sugar molecule are numbered. The 5' carbon is bound to the phosphate group and the 3' carbon is bound to a hydroxyl group (-OH). A nucleotide strand can only grow by bonding the hydroxyl group to the phosphate group, so one end of nucleic acids is always a 5' end and the other is always a 3' end. New nucleotides can only be added to the 3' end of the nucleic-acid strand. Therefore, replication and transcription only occur in the 5' to 3' direction.

Proteins also have directionality. One end of an amino acid contains the NH group and the other end contains the COOH group. This structure gives the amino acids and peptides directionality. A peptide bond can only form between the NH group of one amino acid and the peptide group of another.

Carbohydrates consist of sugars and polymers of sugars. The simplest sugars are monosaccharide, which have the empirical formula of CH_2O. The formula for the monosaccharide glucose, for example, is $C_6H_{12}O_6$. Glucose is an important molecule for cellular respiration, the process of cells extracting energy by breaking bonds through a series of reactions. The individual atoms are then used to rebuild new small molecules. It is important to note that monosaccharides are the smallest functional carbohydrates, so they cannot be hydrolyzed into simpler molecules and still remain carbohydrates.

Polysaccharides are made up of a few hundred to a few thousand monosaccharides linked together. These larger molecules have two major functions. The first is that they can be stored as starches, such as glycogen, and then broken down later for energy. Secondly, they may be used to form strong materials, such as cellulose, which is the firm wall that encloses plant cells, and chitin, the carbohydrate insects use to build exoskeletons.

Cell Organelles Provide Essential Cell Processes

Biological molecules interact with each other to form different structures. The basic unit of biological life is the cell. Eukaryotic cells comprise different subcellular components called organelles, whose structure, function, and interactions help the cell undergo cellular processes. Each organelle has an essential function for sustaining the life of the cell.

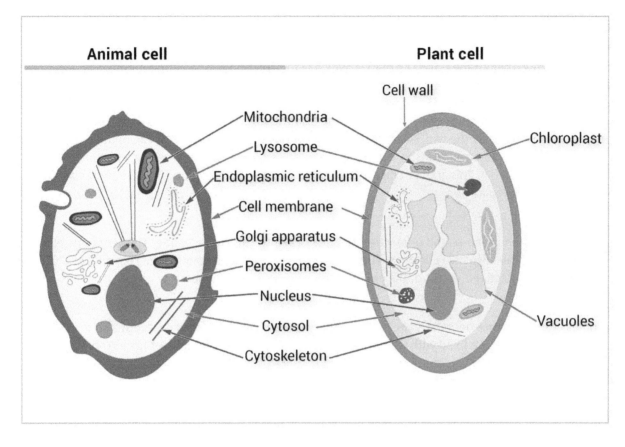

Ribosomes
Ribosomes are made up of ribosomal RNA molecules and a variety of proteins. They are the structures that synthesize proteins. They consist of two subunits, small and large. The ribosomes use messenger RNA as a template for the protein and transfer RNA to bring amino acids to the ribosomes, where they are synthesized into peptide strands using the genetic code provided by the messenger RNA. Most ribosomes are attached to the ER membrane.

Endoplasmic Reticulum (ER)
The ER is a network of membranes contiguous with the outer membrane of the nucleus that forms flat disks. There are two types of ER, rough and smooth. Rough ER is lined with ribosomes and is the location of protein synthesis. This provides a separate compartment for site-specific protein synthesis and is important for the intracellular transport of proteins. Smooth ER does not contain ribosomes and is the location of lipid synthesis.

Golgi Apparatus
The Golgi apparatus is the site where proteins are modified; it is involved in the transport of proteins, lipids, and carbohydrates within the cell. The Golgi apparatus is made up of flat layers of membranes called cisternae. Material is transported in transfer vesicles from the ER to the cis region of the Golgi

apparatus. From there, the material moves through the medial region, is sometimes modified, and then leaves through the trans region of the ER in a secretory vesicle.

Mitochondria

The mitochondrion is the primary site of respiration and adenosine triphosphate (ATP) synthesis inside the cell. Mitochondria have two lipid bilayers that create the intermembrane space, which is the space between the two membranes, and the matrix, which is the space inside the inner membrane. While the outer membrane is smooth, the inner membrane is folded and forms cristae. The outer membrane is permeable to small molecules. The cristae contain many proteins involved in ATP synthesis. In the mitochondria, the products of glycolysis are further oxidized during the citric-acid cycle and the electron-transport chain. The two-layer structure of the mitochondria allows for the buildup of H+ ions produced during the electron-transport chain in the intermembrane space, creating a proton gradient and an energy potential. This gradient drives the formation of ATP. Mitochondria have their own DNA and are capable of replication.

Lysosomes

Lysosomes are membrane-bound organelles that perform intracellular digestion. These organelles have a low pH and contain digestive enzymes. The structure of lysosomes allows for the recycling of cellular material while protecting the rest of the cell from harmful pH and enzymes.

Vacuoles

Vacuoles are membrane-bound organelles found primarily in plant and fungi cells, but also in some animal cells. Vacuoles are filled with water and some enzymes and are important for intracellular digestion and waste removal. The membrane-bound nature of the vacuole allows for the storage of harmful material and poisonous substances. The pressure from the water inside the vacuole also contributes to the structure of plant cells.

Chloroplasts

Chloroplasts are organelles found primarily in plants and are the site of photosynthesis. They have a double membrane and also contain membrane-bound thylakoids, or discs, that are organized into grana, or stacks. Chlorophyll is present in the thylakoids, and the light stage of photosynthesis, which includes the production of ATP and $NADPH_2$, occurs there. Chlorophyll is green and traps the light energy necessary for photosynthesis. Chloroplasts also contain stroma, which are the site of the dark reaction stage of photosynthesis, during which sugar is made. The membrane structure of chloroplasts allows for the compartmentalization of the light and dark stages of photosynthesis.

Differentiation in Development is Triggered by Internal and External Cues

In multicellular organisms, cells work together to perform biological functions. To do this, cells specialize and then form organs. This process occurs early in the development of a multicellular organism. Specialization is determined by internal and external signaling cues that regulate the genes within the cell. As genes are turned on and off, the fate of the cell is determined. For example, in *drosophila*, there are different concentrations of different proteins within certain locations of the egg. These varying concentrations correlate to the eventual structures in that region of the organism. For example, in the posterior region of the embryo, there are higher concentrations of proteins for caudal structures.

Gene Expression Can Alter Which Cells and Tissues Develop

Many of these proteins can bind to DNA and change which genes are expressed. As the embryo develops, the combination of these proteins determines which genes are turned on in each cell, which in turn determines what types of cells and tissues are developed. Mature cells can also respond to external and internal cues by alternating gene expression. This happens in a process known as signal transduction. One example of this process is when a cell is infected by a virus. The cell recognizes the virus and induces a signaling cascade that activates transcription factors, or proteins that can turn genes on and off, to turn on the genes necessary for cell defense.

Specialized organs and tissues must work together for a multicellular organism to function. Organs are organized into organ systems that carry out biological functions. These organ systems must communicate and work together through interactions for optimal function.

Organ Systems Work Together

The urinary-tract system is an organ system that comprises several organs, including the kidney and the bladder. The role of the urinary-tract system is to filter waste and extra fluid from the blood and remove it from the body. The kidneys and the bladder have to work together to complete this process. Throughout the day, the kidneys receive blood and then filter waste products and excess fluid from the blood. The waste produced by the kidneys is urine. As urine is produced, it travels through the ureter and into the bladder, which acts as a holding system until the person is ready to urinate.

Muscle contraction is a voluntary process that requires the nervous system and the musculoskeletal system to function cooperatively. When a person wants to move a muscle, an action potential is sent through the axon of a motor nerve cell, called a motor neuron. Between the motor neuron and the muscle it innervates, there is a gap, or synapse. When the action potential reaches the end of the motor neuron, the neuron releases a neurotransmitter, called acetylcholine, into the synapse. This chemical then signals the muscle and creates an action potential in the muscle, which causes the muscle to contract.

The Structure of Communities

Organisms interact with each other and form complex environments. The structure of these environments comprises:

- Population: A population consists of all of the members of the same species that live in a geographical area and are able to reproduce and create fertile offspring. The size of any given population is determined by many factors, which include birth rate, death rate, and migration.

- Community: A community is made up of all of the different populations that live in a given geographic area. There are complex interactions both within a species and between different species. Communities that are more diverse and more complex are more stable than simple communities.

- Ecosystem: An ecosystem is the basic unit of ecology and combines all of the living, or biotic, organisms and nonliving, or abiotic, components. The abiotic components include light, water, air, minerals, and nitrogen. They provide necessary nutrients and energy for the biotic components. Energy in an ecosystem usually flows from sunlight to primary producers, which are organisms that can undergo photosynthesis, to secondary consumers and finally to decomposers. Abiotic nutrients often have complex cycling systems between different members of an ecosystem.

- Biosphere: A biosphere is the collection of all of the biomes in the world. A biome is a group of ecosystems with similar properties, such as climate and geography. Types of biomes include the tundra, coniferous forests, deserts, lakes, coral reefs, rivers, and savannas.

Population Size

The size of a population is affected by many different factors in a community. Populations grow when new members are born, and they shrink when members die or migrate away. Birth rates, death rates, and migratory rates are affected by factors such as the presence of predators, the availability of food, and the availability of shelter. The population size of predators and prey are linked. The population of a predator grows when the number of prey is great. Eventually, the population of the predator gets so high that competition exists between individuals of the predator species. The population of the prey starts to fall and the competition between the predators gets worse. Eventually, the population of the predator cannot be sustained and the population starts to fall. When the predator population decreases enough, the population of the prey starts to rebound. This relationship can also be affected by external factors. Mathematical modeling can predict this relationship and show the impact of the external factors.

Mathematical Models of Population Growth

Mathematical population models can predict the impact of the community on population growth. In a community where a population has unlimited access to all resources, population growth will be exponential. This is the maximum growth rate that a population can have and is not frequently seen in nature. When the population size exceeds the resources, the growth rate will change because individuals will be competing for resources. Birth rates will slow, while death and migratory rates will rise. The population will enter the logistic growth phase and reach the maximum that the community can carry. Assuming there are no changes to the community, the population will reach a steady state.

The human population growth rate can be predicted by fecundity, or reproductive rate, and the age of the population. For the last 100 years, the global human population growth rate has been very high but, over the past few years, it has started to slow. In certain areas of the world, the growth rate remains high. These places have a young population and a high birth rate. In a young population, there are many people who will still have children and a high birth rate means that a lot of children are born. Other places have a lower birth rate and an older population. In an older population, there are fewer people with reproductive potential and, generally, there is a higher death rate. With a lower birth rate, there are fewer children born to replace the people that die. Almost every part of the world still has an increasing human population growth rate. An exception to this is Japan, where the birth rate is very low and the population is very old and possibly shrinking. Models predict that is may soon happen in Western Europe as well.

The Movement of Matter and Energy

Organisms depend on energy and material for nutrients and survival. Material resources are finite and must circulate within the ecosystem, creating complex recycling systems. Four important resources have well-established cycles:

Water: The sun heats bodies of water and causes evaporation, the changing of water from a liquid to a gas. Plants release water through transpiration, which is when water moves upward through the plant and is then released as a vapor from the leaves into the atmosphere. The cooler temperatures of the atmosphere cause the water to condense back into a liquid, forming a cloud. Eventually, the water in the clouds falls from the sky as precipitation, such as rain, snow, or hail. It then falls back into bodies of water or onto land. On land, precipitation is absorbed into the soil, where it finds its way into bodies of water as runoff or is absorbed into plants.

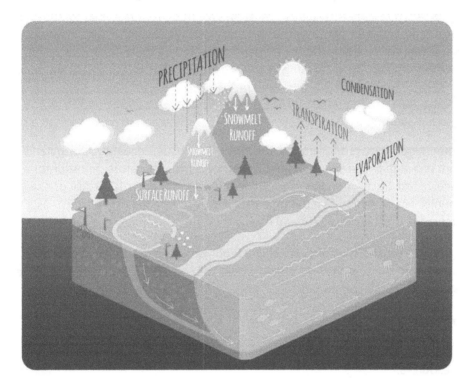

Carbon cycle: Carbon forms the backbone of all biologically important molecules. It is found in the atmosphere as CO_2. Plants, algae, and cyanobacteria take CO_2 and make carbohydrates during photosynthesis using energy from the sun. The carbon then moves through animals in the food chain and is returned to the atmosphere as CO_2 during respiration. Decaying biological material, called *detritus,* also provides carbon to the soil. A final source of carbon is the burning of wood and fossil fuels, which releases CO_2 into the atmosphere.

Nitrogen: Nitrogen is important for protein synthesis. Most nitrogen is found in the atmosphere as N_2, which is unusable to the majority of organisms. Lightening can convert N_2 to nitrates, which can then be used by plants. Nitrogen-fixing bacteria can turn N_2 into nitrates or ammonium. These bacteria can live in the roots of legumes, allowing the nitrates to be consumed as food. Ammonium can also be produced by decomposing material. Ammonium is converted into nitrites and then nitrites are converted to nitrates by nitrifying bacteria. Plants can then take up nitrates by assimilation. Denitrifying bacteria convert nitrates back into N_2.

Phosphorous cycle: Phosphorous is important for plant growth and development. It is a component of RNA, DNA, ATP, cell membranes, and bones; and is an important modifier of proteins. Phosphorous is found in rocks and soil as PO_4^{3-}, which enters the soil from rock erosion or enters lakes from runoff. PO_4^{3-} is then taken up by plants and algae, which can convert the PO_4^{3-} into useable material. Animals receive phosphorous by eating plants. Phosphorous is returned to the soil after the death of animals and plants. Phosphorous can also enter the soil in the form of man-made fertilizer.

Unlike these recyclable resources, energy flows from sunlight to primary producers to consumers and is not recycled. Between each step, there is actually a loss of energy to the environment, biological processes, and heat generation. Primary producers are the first step in harnessing the energy from sunlight by creating glucose from CO_2 and sunlight. This reaction is greatly affected by changes to the environment. Since light is a necessary factor for photosynthesis, the seasonal variation in daylight affects the rate of photosynthesis. Temperature is also a factor. Photosynthesis slows at low temperatures and transpiration increases at high temperatures, creating water loss. Under drought conditions or high temperatures, photosynthesis occurs less. Photosynthesis increases as the levels of CO_2 increase. Since rates of photosynthesis are greatly affected by these resources, climate change has an impact on photosynthesis. The increasing CO_2 levels impact photosynthesis, as do drastic changes in weather patterns, such as temperature shifts and droughts.

Food Webs

Energy flows from the primary producers to consumers, as consumers eat producers and other consumers. From one step to the next, 90% of the energy is lost. Therefore, in order for a tertiary consumer to receive just 1 kcal of energy, a producer needs to produce 1000 kcal of energy. The energy flow in a community can be shown as a pyramid that illustrates the loss of energy or as a web that illustrates the complex relationships between each organism.

It is important to note that the start of every single food web has a producer. Without a producer to harness the energy from the sun, the entire web would collapse. There would not be any energy to sustain any populations in the community.

Models of Competition

Mathematical modeling can demonstrate how changes in both biotic and abiotic factors affect the use of energy and resources.

Competition for resources is best modeled by the logistic model of population growth. Competition for resources, territoriality, health, predation, and accumulation of waste regulate the density at which a population can survive. When the population is too dense, there are not enough resources, so disease spreads and waste accumulates.

Human Activities Impact Ecosystems

Humans often have the biggest impact on ecosystems. Our population size and activities often disrupt the ecosystems of other organisms. Among the most impactful activities of humans on ecosystems is deforestation. Forests are torn down for lumber to create homes for our growing population, as well as to create farmland to feed our growing population. This deforestation destroys the habitats of many organisms and can lead to their death. Many species have become extinct because of deforestation and loss of habitat. Examples of species that are extinct because of deforestation include the Tasmanian tiger, the passenger pigeon, and the Javan tiger. Pandas, orangutans, tigers, and gorillas are among the many species in danger of extinction from deforestation. The larger the human population gets, the more their activities impact ecosystems. What's more, this growing impact occurs at a rate disproportionately higher than the rate of increase of the human population.

Adaptations are Often Due to the Ability to Use Energy

The ability to gain and conserve energy is one of the greatest selective qualities. Therefore, many species have adaptations that help them use energy and matter more efficiently. One example of this is the adaptation of Darwin's finches. Different species of these finches have different shaped bills that can access different sources of food more efficiently.

Changes in Structure Can Cause Changes in Function

Interactions between molecules are the driving force behind biological processes. The structures of the different molecules facilitate their interaction. Biological processes progress because this interaction causes a change in the structure of one or more of the molecules, facilitating interactions with new molecules.

The Effects of Enzyme Shapes

Enzymes are proteins that catalyze a reaction, which means they make a reaction happen faster. Almost all biological reactions are catalyzed by enzymes. In order for the enzyme to function, it needs to interact with the substrates, which are the components of the reaction. This process is very precise to ensure that only the correct substrates are involved in the reaction. Therefore, the substrates have a very specific fit into the active site of the enzyme, which is the area that catalyzes the reaction. Once the enzyme and substrate bind to the active site, they form what is called the *enzyme/substrate complex.*

Cofactors and Coenzymes

To function correctly, some enzymes require cofactors, such as vitamins, and coenzymes. These molecules affect the shape of the enzyme. When they are not present, the substrate cannot fit into and interact with the enzyme. However, when the cofactors or coenzymes are present, the shape of the enzyme changes, and it can interact with the substrate.

Because this fit needs to be so perfect, it offers a method for regulation of enzyme activity. Other factors or molecules may bind to the enzyme and change its shape. This alters the enzyme's ability to bind to the substrates. These molecules bind to allosteric sites, which are sites that are different from the active site. This binding can either change the shape to allow the substrate to fit, which is called allosteric activation, or it can prohibit the substrate from fitting, which is called allosteric inhibition.

Specialization of Cellular Organelles and Structures

Being able to use energy efficiently benefits organisms. To do this, biological activities are performed by specialized mechanisms, both at the cellular and the organismal level. This specialization allows each compartment to excel at its role and not waste energy on unrelated tasks.

Within a cell, the plasma membrane transports the material necessary for biological functions, the cytoplasm provides an aqueous environment for cellular activities, and the different organelles each perform particular tasks that help the cell function.

Cell membrane: All cells are surrounded by a cell membrane that is made up of lipids and proteins. The cell membrane is created from two layers of phospholipids, which are two fatty-acid chains connected to a glycerol molecule that is connected to a phosphate group. The fatty-acid chains are hydrophobic, while the phosphate group is hydrophilic, which makes the entire phospholipid amphiphilic. Since the outside of a cell, known as the extracellular space, and the inside of a cell, the intercellular space, are aqueous, the lipid bilayer forms with the two phospholipid heads facing the outside and the inside of the cell. In this way, the phospholipids can interact with water and the fatty-acid tails, which face the middle, can interact with each other and avoid water.

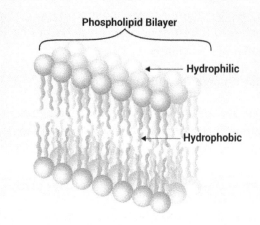

The lipid that makes up the cell membrane creates a unique environment that protects the cell from the outside environment and allows material to pass through the membrane. This property is called semipermeability. Some molecules can pass through the membrane, while others need to use protein channels. This allows the cell to regulate the passage of certain materials.

Specialization of Organs

Specialization also happens within organs and organ systems, which allow the cells in that organ to turn off genes and processes not related to that function, thereby conserving energy. For example, the digestive system carries out the process of breaking down food so that nutrients can be taken up by cells. The mouth chews the food and creates saliva with the enzymes necessary for the breakdown of some starches. Swallowing brings the food into the esophagus and down to the stomach, where acidic gastric juices are secreted that break down food further and kill any bacteria present in the food. From the stomach, food travels to the small intestines, where the food is broken down into basic molecular units and absorbed through the walls of the intestines into the blood stream. Waste continues on through the large intestines and is eliminated from the body as feces.

Community Specialization

While many single-cellular organisms do not have the ability to specialize, some live in communities where they take on specialized roles. These communities have increased energy consumption and efficiency. One example of this is the bacteria that live in the guts of ruminants, including cattle, goats, sheep, and deer, which have a diet that consists of plant material. To obtain the energy from the cellulose, these animals have bacteria in their stomachs that can break down the cellulose. The bacteria provide a specialized function for the animal and receive a food source in return, increasing the energy efficiency of both organisms.

Interactions Between Populations

Different populations interact to create the complex functions of a community. These interactions can have both positive and negative impacts on the individuals involved from the different populations and can be modeled mathematically. The models can demonstrate how the negative or positive relationship will impact the population of each species in the relationship. There are five types of interactions:

Competition: Competition is when two individuals vie for a finite amount of resources, such as food, water, and mates. This can occur within or between species. Both groups are negatively affected by competition.

Predation: Predation occurs when one species, the predator, feeds on another species, the prey. Predation usually, but not always, ends in the death of the prey. This relationship is beneficial to the predator and harmful to the prey. Ultimately though, as the number of prey decrease (from predation), the number of predators eventually decrease, as food supplies dwindle.

Parasitism: Parasitism is another relationship where one species, the parasite, gains a benefit from the relationship, but the other, the host, is harmed by the relationship. Unlike predation, however, parasitism does not always result in the death of the host and does not always involve a way to ascertain food. Parasites can use their host as a food source, but they can also use their hosts as a place to lay their eggs for reproduction or to provide a habitat.

Commensalism: Commensalism is when one member of the relationship benefits and the other is not affected at all. Examples of this include when one organism uses another for transport or housing without harming the other organism.

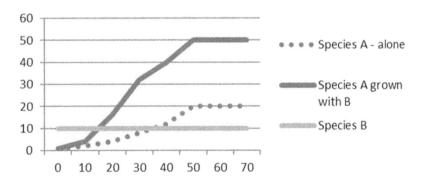

Mutualism: Mutualism is a relationship where both members benefit. There are many examples of this, including: the nectar-drinking/pollination relationship between insects or birds and plants; animals eating fruit and dispersing seeds; and animals that feed on parasites on other animals.

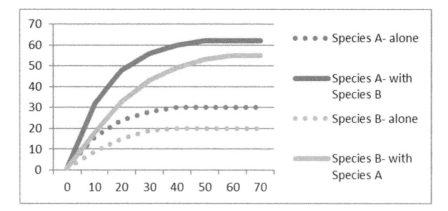

All of these relationships occur in the context of an entire ecosystem. They are not solitary relationships and are affected greatly by other forces outside of the relationship. This allows for feedback mechanisms to control the symbiotic relationships.

Population Dynamics

Population dynamics can be influenced by many factors, both biotic and abiotic. Natural and man-made disasters can greatly affect population size and species distribution. One example of this is the destruction of elm trees by Dutch elm disease. This fungus is spread by beetles and was accidentally introduced to Europe, North American, and New Zealand. The disease is native to Asia and many trees there are resistant to the disease. Dutch elm disease wiped out approximately 75% of the elm trees through much of Europe.

Distribution of Ecosystems

While ecosystems are often discussed in their current state, it is important to note that ecosystems are not static; they change over time, which can be a natural process. Weather patterns, geological events, and fires are among the things that can affect an ecosystem. These items can reduce or increase resources, destroy habitats, or even kill individuals. One example of a natural change to an ecosystem is the weather pattern known as El Niño, which is caused by increased water temperatures in the Pacific Ocean and results in weather changes throughout the world. El Niño can have an effect on many ecosystems. In the ocean off the coast of Peru, there are fewer predatory fish because the ocean conditions provide fewer nutrients for the plankton that the fish eat. In areas that experience flooding from El Niño, there can be overflow of salt water into fresh-water systems, which is destructive to those ecosystems. Areas with droughts can see a loss of producers, which is also destructive to ecosystems.

Human Impact Accelerates Change

Humans can cause change to an ecosystem at an even faster rate than nature. Humans destroy habitats and introduce species that have the potential to be very invasive to new areas. One example of this is the introduction of smallpox to the Western Hemisphere. Smallpox was a disease caused by a virus that had existed in the Eastern Hemisphere since around 10,000 BC. It did not exist in the Western Hemisphere until European exploration in the fifteenth century. The European travelers exposed the native populations of the Western Hemisphere to smallpox. Since this population had no existing immunity, the population was decimated.

Molecular Diversity

Diversity exists in biology, even among items that are the same or similar, such as molecules, cells, and species. This diversity allows for a wider range of function and a greater ability to respond to the environment. One example of this is the production of antibodies by plasma cells. Antibodies are a class of proteins that provides a defense mechanism against pathogens. Their general structure is the same but each antibody has a slight variation at its tip that differentiates it for its specific pathogen. The plasma cells that make antibodies can make small changes in the antibody's DNA sequence to allow for the production of a large, diverse population of antibodies.

Organisms also have molecular diversity through genetic variation. One mechanism for creating genetic variation is by having two or more alternative forms, or alleles, of the same gene, such as with eye color or hair color. If both alleles are identical, the individual is considered homozygous; if the two alleles have different sequences, the individual is considered heterozygous. In most genes, one allele is considered more dominant than the other and will mask the appearance of the less dominant, or recessive, allele when there is a heterozygous situation. Another type of genetic variation that occurs is through the introduction of a mutation, which is a random, permanent alteration of a gene sequence. Some mutations that occur can be advantageous to a population and may help a population adapt to its environment. This may allow the organism to respond to stresses with more resilience and may even give them a fitness advantage. One example of a fitness advantage is in people whose hemoglobin gene is heterozygous for the sickle-cell trait. People who are homozygous for the sickle-cell trait have sickle-cell anemia, a harmful genetic disorder. People who are heterozygous for the sickle-cell trait are more resistant to malaria than people who are homozygous for wild-type, or normal, hemoglobin.

Gene Duplication

Genetic variation can also be established if a gene duplicates. Now the organism has an extra set of this gene, which gives the gene more chances to mutate since the original copies still maintain the necessary function of that gene. One example of this is the development of the antifreeze genes in fish that live in arctic environments, which prevents them from freezing. Originally, these fish had one copy of the sialic-acid synthase (SAS)–B gene. That gene was duplicated and, since the original SAS-B gene still retained its necessary function, the duplicated gene was no longer under selective pressure and could mutate. It then developed the mutations necessary to prevent fatal ice crystals from growing inside the fish.

Environmental Factors Influence Traits

Biological variation can also exist from environmental factors influencing gene expression. Two organisms with identical genomes can have different genes turned on or off depending on their different environments. The effect of the presence of lactose on bacteria that are Lac+ is one example. Lac+ bacteria have a set of genes that metabolize the sugar lactose. When these bacteria are not around lactose, a repressor protein stops transcription of these genes. However, the presence of lactose causes the degradation of the repressor and the genes that metabolize lactose are transcribed. These environmental influences can also be seen at the organism level. Some cats have a temperature-sensitive mutation in a gene that is important for making fur color. At normal body temperature, this mutation disables the gene's function, so the fur is a light color. The cat's extremities are cooler and at a lower temperature, so the gene functions, giving the cat dark fur on its extremities.

Populations Need to Respond to Environmental Changes to Survive

The accumulation of these potential genetic variations can result in a diverse population. Populations with a high level of genetic diversity are able to respond better to changes in the environment, while populations with little genetic diversity are at risk for extinction.

One example of this is the destruction of potato crops in the mid-nineteenth century in Ireland because of a potato blight, which is a fungal infection. This destruction caused widespread famine and the death of hundreds of thousands of people. Potatoes can be grown by vegetative propagation, a type of asexual reproduction. The result of this asexual reproduction technique was that most of the potatoes in Ireland were genetic clones. All of these potatoes were highly susceptible to potato blight. If different strains of potatoes had been planted in Ireland, they would have had different levels of resistance to potato blight, and the famine caused by the loss of the potato plants would not have been as destructive.

Genetic diversity allows individuals in a population to respond differently to the same stimuli. This is seen with how individuals respond during a disease outbreak. Some individuals succumb to the disease, some are made ill, and others are not infected at all. These different responses are caused by the genetic makeup of the individual. Some individuals have genes that will prevent infection completely, others have immunological genes that can fight off infections, and still others have immunological genes that are unable to fight off the infection at all.

Allelic Variation Can be Modeled by the Hardy- Weinberg Equation
The different alleles that are present in a population can be modeled by the Hardy-Weinberg equilibrium. The Hardy-Weinberg equilibrium states that, absent any evolutionary influence, the proportion of alleles will remain constant throughout different generations. In cases where there are two different alleles, this equilibrium can be explained with the equation $p^2+2pq+q^2=1$, where p is the proportion of one allele in the population and q is the proportion of the other allele. In this equation, p^2 represents the number of homozygotes of one allele, 2pq represents the number of heterozygotes, and q^2 represents the number of homozygotes for the other allele.

Ecosystem Diversity

Ecosystem diversity is represented by the number of different species in the ecosystem. Ecosystems with higher diversity are more resilient to changes in the environment than simple ecosystems because the loss of any one species is not as detrimental.

The key factors in maintaining diversity in an ecosystem are keystone species, producers, and essential biotic and abiotic factors. Keystone species are any species whose role in the ecosystem is disproportionate to the size of the population. Although the keystone species may not be the most numerous or the most productive part of an ecosystem, its loss would devastate the ecosystem. For example, a keystone species may be a small predator that preys on an herbivorous species and keeps that species from eliminating all of a particular plant species. If the keystone species became extinct, the herbivorous species would completely wipe out the plant species and the ecosystem would change drastically. Similarly, in large bodies of water, the sea star is a keystone species that preys on sea urchins, which helps protect the coral reefs. Within an ecosystem, each species plays a specific, important role in preserving the environment they populate together.

Practice Questions

1. How do cellulose and starch differ?
   a. Cellulose and starch are proteins with different R groups.
   b. Cellulose is a polysaccharide made up of glucose molecules and starch is a polysaccharide made up of galactose molecules.
   c. Cellulose and starch are both polysaccharides made up of glucose molecules, but they are connected with different types of bonds.
   d. Cellulose and starch are the same molecule, but cellulose is made by plants and starch is made by animals.

2. A mutation in the sequence of a protein causes the secondary structure to change. How did the mutation cause the change?
   a. The R group from the mutated amino acid interacts differently with other R groups.
   b. The R group from the mutated amino acid prevents the formation of hydrogen bonds between the atoms of the backbone of the protein.
   c. The mutation causes peptide bonds to change.
   d. All of the above

3. Palmitoleic acid is a fatty acid with one double bond in the hydrocarbon chain. What property would you expect from palmitoleic acid?
   a. Solid at room temperature
   b. Gas at room temperature
   c. Liquid at room temperature
   d. Hydrophilic

4. What molecule serves as the hereditary material for prokaryotic and eukaryotic cells?
   a. Proteins
   b. Carbohydrates
   c. Lipids
   d. DNA

5. What organelles have two layers of membranes?
   a. Nucleus, chloroplast, mitochondria
   b. Nucleus, Gogli apparatus, mitochondria
   c. ER, chloroplast, lysosome
   d. Chloroplast, lysosome, ER

6. What organelle is the site of protein synthesis?
   a. Nucleus
   b. Smooth ER
   c. Ribosome
   d. Lysosome

7. What happens to the population of a predator if the population of a prey decreases?
    a. Increases
    b. Decreases
    c. Stay the same
    d. Not enough information to answer the question

8. If an ecosystem lost its denitrifying bacteria, where and in what form would nitrogen accumulate?
    a. In the soil as nitrates
    b. In the air as N_2
    c. In the soil as ammonium
    d. In the soil as N_2

9. An ecosystem that normally has moderate summers with high rainfall is experiencing a heat wave and a drought. How does this affect the rate of photosynthesis of the producers in this ecosystem?
    a. The increase in transpiration from the high heat and the drop in rainfall result in less water available for photosynthesis. The rate decreases.
    b. The increase in transpiration from the high heat and the drop in rainfall result in less water. Since photosynthesis creates water, the rate increases to meet increased water demands.
    c. Water availability has no effect on photosynthesis.
    d. Increased temperature increases the number of mitochondria, so photosynthesis rates increase.

10. What is the role of an allosteric activator?
    a. To bind to the active site of an enzyme and block the binding of a substrate.
    b. To bind to the active site of an enzyme and allow the binding of a substrate.
    c. To bind to an unrelated site of an enzyme to block the binding of a substrate.
    d. To bind to an unrelated site of an enzyme to allow the binding of a substrate.

11. Certain bacteria that can break down the bonds in cellulose live in the gut of ruminants, which are mammals that feed primarily on grasses. Animals cannot break down cellulose. How does this affect the energy efficiency of both the bacteria and the ruminants?
    a. Energy efficiency of the bacteria increases. Energy efficiency of the ruminants decreases.
    b. Energy efficiency of the bacteria decreases. Energy efficiency of the ruminants decreases.
    c. Energy efficiency of the bacteria decreases. Energy efficiency of the ruminants increases.
    d. Energy efficiency of the bacteria increases. Energy efficiency of the ruminants increases.

12. A new species is introduced into an ecosystem. This species is a parasite to an existing species in the ecosystem, the host. What will the immediate effects be on the population size of the parasite and the host?
    a. Parasite increases, host decreases
    b. Parasite increases, host increases
    c. Parasite decreases, host increases
    d. Parasite decreases, host decreases

13. A farmer grows all of his tomato plants by vegetative propagation. He finds that one clone produces tomatoes that sell much better than any other clone. He then uses this clone to plant his entire field. Two years later a fungus wipes out his entire crop. What could the farmer have done to prevent this?
    a. Plant tomatoes that sell poorly. They are more resistant to fungus.
    b. Plant a variety of tomatoes. Genetic variation would have left some of the crop less susceptible to the fungus.
    c. Plant a variety of crops. Plants other than tomatoes might not be affected by the fungus.
    d. B or C

14. Two different bacterial cultures are grown from bacteria with the same genome sequence. Transcriptional analysis shows that Culture B is expressing genes that can metabolize lactose, but Culture A is not. How can this happen if they have the same genetic sequence?
    a. Someone mislabeled the tubes and the bacteria must have different genome sequences.
    b. Culture A is grown in the presence of lactose, which turns on a different set of genes.
    c. Culture B is grown in the presence of lactose, which turns on a different set of genes.
    d. B and C

15. An ecosystem experiences a loss of one species due to hunting. While the overall population size of this species was small, the loss of this species is devastating to the ecosystem. What kind of species was this?
    a. Producer
    b. Consumer
    c. Parasite
    d. Keystone

Answer Explanations

1. C: Cellulose and starch are both polysaccharides that are long chains of glucose molecules, but they are connected by different types of bonds, which gives them different structures and different functions.

2. B: Secondary structure is formed from hydrogen bonds between the backbone atoms in the protein chain. Some R groups allow these interactions to form, and others prevent them. A mutation can cause the secondary structure to change when it changes an R group that allows these interactions to one that prevents these interactions.

3. C: Palmitoleic acid is an unsaturated fatty acid. They are typically liquids at room temperature. They are also hydrophobic.

4. D: DNA serves as the hereditary material for prokaryotic and eukaryotic cells.

5. A: The nucleus, chloroplast, and mitochondria are all bound by two layers of membrane. The Golgi apparatus, lysosome, and ER only have one membrane layer.

6. C: Proteins are synthesized on ribosomes. The ribosome uses messenger RNA as a template and transfer RNA brings amino acids to the ribosome where they are synthesized into peptide strands using the genetic code provided by the messenger RNA.

7. B: When the population of a prey decreases, the population of the predator will also decrease as competition increases between the individuals in the predator population and as the prey resource becomes scarce.

8. A: Denitrifying bacteria live in the soil and convert nitrates to N_2. If they did not exist, nitrates would accumulate in the soil.

9. A: Water is essential for photosynthesis. Increasing temperatures increase transpiration and drought conditions result in less water available for photosynthesis. The rate of photosynthesis will decrease.

10. D: Allosteric activators bind to an allosteric site, a site other than the active site, of an enzyme and cause a conformational change that allows the substrate to bind to the active site of the enzyme.

11. D: The ruminants provide a food sources for the bacteria, and the bacteria help the ruminants utilize their main food source. Therefore, the energy efficiency of both organisms increases.

12. A: A parasite benefits from the relationship with the host, while the host suffers a fitness cost. Therefore, the population of the parasite will increase while the population of the host will decrease.

13. D: Genetic variety in a species allows them to be more resistant to stresses. Having genetic diversity increases resilience. Growing multiple strains of tomatoes or multiple types of crops could protect the farm.

14. C: Gene expression can be influenced by the environment. Lactose metabolism is regulated by the presence of lactose. Bacteria that have the genes to metabolize lactose will turn them off if lactose is not present but will turn them on if lactose is present.

15. D: Keystone species are any species whose role in the ecosystem is disproportionate to the size of the population. The loss of a keystone species will devastate an ecosystem.

Dear AP Biology Test Taker,

We would like to start by thanking you for purchasing this study guide for your AP Biology exam. We hope that we exceeded your expectations.

Our goal in creating this study guide was to cover all of the topics that you will see on the test. We also strove to make our practice questions as similar as possible to what you will encounter on test day. With that being said, if you found something that you feel was not up to your standards, please send us an email and let us know.

We would also like to let you know about other books in our catalog that may interest you.

AP Comparative Government and Politics

This can be found on Amazon: amazon.com/dp/1628456167

SAT Math 1

amazon.com/dp/1628454717

SAT

amazon.com/dp/1628454679

ACT

amazon.com/dp/1628454709

ACCUPLACER

amazon.com/dp/162845492X

We have study guides in a wide variety of fields. If the one you are looking for isn't listed above, then try searching for it on Amazon or send us an email.

Thanks Again and Happy Testing!
Product Development Team
info@studyguideteam.com

Interested in buying more than 10 copies of our product? Contact us about bulk discounts:

bulkorders@studyguideteam.com

FREE Test Taking Tips DVD Offer

To help us better serve you, we have developed a Test Taking Tips DVD that we would like to give you for FREE. **This DVD covers world-class test taking tips that you can use to be even more successful when you are taking your test.**

All that we ask is that you email us your feedback about your study guide. Please let us know what you thought about it – whether that is good, bad or indifferent.

To get your **FREE Test Taking Tips DVD**, email freedvd@studyguideteam.com with "FREE DVD" in the subject line and the following information in the body of the email:

    a. The title of your study guide.

    b. Your product rating on a scale of 1-5, with 5 being the highest rating.

    c. Your feedback about the study guide. What did you think of it?

    d. Your full name and shipping address to send your free DVD.

If you have any questions or concerns, please don't hesitate to contact us at freedvd@studyguideteam.com.

Thanks again!

Lightning Source UK Ltd.
Milton Keynes UK
UKHW031155080720
366212UK00007B/187

9 781628 456226